Pioneers in Arts, Humanities, Science, Engineering, Practice

Volume 3

Series editor

Hans Günter Brauch, Mosbach, Germany

More information about this series at http://www.springer.com/series/15230
http://www.afes-press-books.de/html/PAHSEP.htm
http://afes-press-books.de/html/PAHSEP_Lebow.htm

Richard Ned Lebow
Editor

Richard Ned Lebow: Major Texts on Methods and Philosophy of Science

 Springer

Editor
Richard Ned Lebow
Department of War Studies
King's College London
London
UK

Acknowledgement: The cover photograph was taken from the author's honorary degree ceremony in Athens (Greece). All other photos in this volume were taken from the personal photo collection of the author who also granted the permission on their publication in this volume. A book website with additional information on Richard Ned Lebow, including videos and his major book covers is at: http://afes-press-books.de/html/PAHSEP_Lebow.htm.

ISSN 2509-5579   ISSN 2509-5587   (electronic)
Pioneers in Arts, Humanities, Science, Engineering, Practice
ISBN 978-3-319-82018-7   ISBN 978-3-319-40027-3   (eBook)
DOI 10.1007/978-3-319-40027-3

Copyediting: PD Dr. Hans Günter Brauch, AFES-PRESS e.V., Mosbach, Germany

Printed on acid-free paper

This Springer imprint is published by Springer Nature
The registered company is Springer International Publishing AG Switzerland

To Lola and Mervyn Frost,
good friends and fellow spirits

Carol Bohmer. *Source* From the author's personal photo collection

Acknowledgements

I would like to thank Hans Günter Brauch again for making this project and volume possible.

Etna, New Hamsphire   Richard Ned Lebow
July 2015

Contents

Chapter 1
Introduction

Richard Ned Lebow

In the early 1960s, Yale political scientists sought to turn its students into shock troops for the behavioral revolution. While sympathetic to learning general lessons about political behavior I rebelled against the crudeness of contemporary theoretical formulations and their underlying epistemological assumptions. Their two most questionable assumptions were that their concepts could accurately describe behavior across cultures and epochs and that correlational analysis would promote knowledge in the form of enhanced ability to predict important outcomes.

My work in philosophy of science has attempted to develop alternate conceptions of knowledge and methods appropriate to them. This began with my turn to case studies to probe the causes of war and develop my critique of deterrence as a strategy of conflict management.[1] I used these and other case studies to explore the relationship between the general and the particular, and highlight the ways in which context—agency, path dependence, confluence, learning, domestic politics and complex agendas—play determinate roles in individual events.

Mark Lichbach and I edited a volume on theory and evidence in comparative politics and international relations.[2] My opening essay, in Chap. 2, develops a critique of positivist approaches to evidence and inference, and poses a series of questions about what constitutes evidence and how it can be used. In lieu of general laws and prediction as the goals of social science, I urge us to reframe our enterprise

---

[1]Most notably, *Between Peace and War: The Nature of International Crisis* (Baltimore: Johns Hopkins University Press, 18981); Robert Jervis, Richard Ned Lebow and Janice Gross Stein, *The Psychology of Deterrence* (Baltimore: Johns Hopkins University Press, 18984); Richard Ned Lebow and Janice Gross Stein, *We All Lost the Cold War* (Princeton: Princeton University Press, 1994).

[2]Richard Ned Lebow and Mark I. Lichbach, *Theory and Evidence in Comparative Politics and International Relations* (New York: Palgrave-Macmillan, 2007).

© The Author(s) 2016
R.N. Lebow (ed.), *Richard Ned Lebow: Major Texts on Methods and Philosophy of Science*, Pioneers in Arts, Humanities, Science, Engineering, Practice 3, DOI 10.1007/978-3-319-40027-3_1

as a practical art along the lines of clinical medicine. We can use any general understanding we like as a starting point for a forecast. It takes the form of a narrative, or multiple narratives that lead to envisaged futures, which branching points and identification beforehand of the kind of information that would enhance one's confidence in these or alternative story lines. Predicting the future is almost impossible, but getting early warning that one's expectations are wrong is feasible and often very helpful.

Forbidden Fruit: Counterfactuals and International Relations (2010) continues my investigation of this subject, as does *Archduke Franz Ferdinand Lives! A World Without World War I* (2014). I use counterfactuals to probe the contingency of World War I and the end of the Cold War and their non-linear causation. They make use of the protocols I developed for conducting more robust counterfactual experiments. I conduct experiments to explore why policymakers, historians and international relations scholars are often resistant to the contingency inherent in open-ended, non-linear systems. Most controversially, I argue that the difference between counterfactual and so-called factual arguments is more one of degree than of kind. The use of counterfactuals by scholars and novelists also challenges the binary between fact and fiction. This volume includes my counterfactual study of World War I and a short story about the possible political consequences of Mozart having lived to the age of sixty five. I used the latter—chapter seven of *Forbidden Fruit*—as instrument in an experiment to probe the relationship between the vividness and credibility of counterfactuals.

More recently, I have worked on the problem of causation in international relations.[3] Cause is a problematic concept in social science, as in all fields of knowledge. We organize information in terms of cause and effect to impose order on the world, but this can impede a more sophisticated analysis. I review understandings of cause in physics and philosophy and conclude that no formulation is logically defensible and universal in its coverage. This is because cause is not a feature of the world but a cognitive shorthand we use to make sense of it. In practice, causal inference is always rhetorical and must accordingly be judged on grounds of practicality. I develop a new—inefficient causation—that is constructivist in its emphasis on the reasons people have for acting as they do, but turns to other approaches to understand the aggregation of their behavior. It is a framework for combining general understandings of political behavior with idiosyncratic features of context.

My most recent work in philosophy of science is a book on Max Weber and International Relations.[4] My chapters in this edited volume address Weber's epistemology, his political views, the relationship between the two and their implications for international relations theory. The last selection is my chapter on Max Weber and knowledge.

---

[3] Richard Ned Lebow, *Constructing Cause in International Relations* (Cambridge: Cambridge University Press, 2014).

[4] Richard Ned Lebow, ed., *Max Weber and International Relations*, forthcoming.

Chapter 2
What Can We Know? How Do We Know?

Richard Ned Lebow

2.1 Introduction

This book[1] was conceived in the course of a long, wet afternoon in Columbus, Ohio. Inside, in a small, brightly lit auditorium, enthusiastic graduate students took turns presenting papers that were the product of a year-long seminar intended to help them develop dissertation proposals. Their words fell on the ears of their fellow students and six professors in international relations. Their presentations, although diverse in subject, were remarkably uniform in structure. They began by laying out a few propositions, went on to describe the data sets or cases that would be used to test these propositions and ended with a discussion of preliminary research findings. The professor who had taught the student participants exuded an avuncular aura throughout the proceedings, and my colleagues, who were encouraged to interrogate the students, largely queried them about their research design and choice of data. For the most part, the students provided competent answers to these questions.

Another colleague and I raised the tension in the room by asking each of the students in turn why they had been drawn to their subject matter. What puzzle or policy concern animated them? What light might their preliminary findings throw on that puzzle or problem? Their responses were largely unsatisfactory. Two students were flummoxed. One insisted he was "filling a gap in the literature." Two

[1] This text was first published as: What Can We Know? How Do We Know," in Richard Ned Lebow and Mark Lichbach, eds., *Political Knowledge and Social Inquiry* (New York: Palgrave, 2007), pp. 1–22, ISBN 9781403974563. The permission to republish this text was granted on 18 June 2015 by Claire Smith, Senior Rights Assistant, Nature Publishing Group & Palgrave Macmillan, London, UK.

R.N. Lebow (ed.), *Richard Ned Lebow: Major Texts on Methods and Philosophy of Science*, Pioneers in Arts, Humanities, Science, Engineering, Practice 3, DOI 10.1007/978-3-319-40027-3_2

more defended their choices in terms of the availability of data sets. Another noted that his subject was a "hot topic," and that a dissertation on it would increase his chances of landing a good job. Only one student justified her research with reference to her sense of urgency about a real world problem: regional conflict.

When pushed, she nevertheless found it difficult to describe what implications her propositions might have for the trajectory of these conflicts or the efforts to ameliorate them. Another colleague, also dissatisfied, questioned the choice of two of the data sets, suggesting that they lumped together cases that had played out in quite different political—historical contexts. After the session, two of my colleagues, including the professor in charge of the seminar, told me I had been too hard on the students.

Two other colleagues were supportive, one of whom, from another field, had heard reports about what had transpired. The three of us agreed that our students, beginning in their introductory scope and methods class, were encouraged to privilege quantitative over qualitative research and choose dissertation topics based more on their feasibility than on their theoretical or substantive importance. They had a sophisticated understanding of research design—but only in so far as it pertained to the strictures of statistical inference. Despite—or perhaps, because—of three years of graduate training, they were correspondingly uninformed about the more general problems concerning evidence. Most gave the impression that it was just "out there" waiting for them to harvest, and failed to realize the extent to which it is an artifact of their theories. They were largely insensitive to context and the understandings of the actors, and how they might determine the meaning of whatever observations they as researchers made. All their proposals conveyed a narrow understanding of science as a form of inference whose ultimate goal is predictive theories. They were not particularly interested in causal mechanisms, let alone in other forms of political understanding such as the constitution of actors.

We agreed that epistemological and methodological narrowness, although pronounced at Ohio State University, was common enough in the discipline to arouse general concern. In our view, the use of King, Keohane, and Verba (KKV), *Designing Social Inquiry*, as a core reading in so many scope and methods courses could only make the situation worse. My colleague, whose reputation was based on 'mainstream' quantitative research—a shorthand term I use to describe those who more or less accept the unity of the sciences—felt just as strongly as I did. He considered many of KKV's recommendations for collecting and evaluating data quite sensible, but he rejected its epistemological foundations as seriously flawed, its characterization of science as ill-informed, relegation of qualitative research to second-class status as unacceptable, and its almost exclusive focus on the construction and analysis of data sets as regrettably narrow. Conversations with a few other dissatisfied colleagues at Ohio State and other institutions led us to consider a book to address some of these concerns.

We did not want to produce another text, nor a study that sanctioned a particular approach. Our goal was to encourage dialogue in the discipline, and among our students, to transcend epistemological and methodological differences. We must pursue our quest for political knowledge as equals because none of our preferred epistemologies are problem free—quite the reverse. Despite inflated claims by partisans of particular approaches, none of them can point to a string of unalloyed theoretical and empirical triumphs that rightfully leave adherents of other approaches frustrated and envious. We can all benefit from a more thorough understanding of each other's assumptions, strategies, practices, successes and failures, and reasons for pride and self-doubt. Such comparison reveals that many of the epistemological and methodological problems we face cut across approaches and fields of study.

With this end in mind, we commissioned representatives of three different epistemologies to write papers on how evidence matters or should matter in the social sciences. These papers were presented and discussed at a conference at Ohio State, hosted by its Mershon Center, on May 12–13, 2000.[2] Some of the papers were revised and presented, with additional ones, at the September 2000 annual scientific meeting of the *American Political Science Association* (APSA). Our book includes some of these papers as well as others that were subsequently commissioned. The conference and APSA panel were characterized by sharp disagreements among people from different research traditions. They also witnessed—as do the succeeding chapters—serious efforts at mutual engagement in the context of addressing problems of common concern. We hope readers will find this tension refreshing and informative.

Our choice of evidence as the initial focus of our papers reflected our commitment to dialogue. Most of us take evidence seriously, recognize that it comes in many forms, and want to develop and apply good procedures for its selection and evaluation. We recognize that our procedures and protocols are far from being problem free and that our treatment of evidence in practice never quite measures up to our ideals. While the papers and subsequent chapters all address the question of evidence, they also speak to problems of epistemology and ontology because evidence cannot satisfactorily be addressed in a philosophical vacuum. The purposes for which we seek and use evidence influence—if not determine—the kind of evidence we seek and the procedures we use to collect, evaluate, and analyze it. Our purposes, in turn, reflect our understandings of the nature of knowledge and how it is obtained. Such assumptions are often left implicit; they may be only partially formulated. All the more reason then to foreground these choices and some of their most important implications for research.

Essays of this kind are messier, make more demands on readers, and inevitably raise more questions than they answer. This is a fair price to pay because the alternative—an effort to "get on with the job" by focusing exclusively, or nearly exclusively, on research methods—clearly the message of KKV—risks missing the forest for the trees. Like KKV, it is likely to conceive of research design in a

---

[2]In attendance were Steven Bernstein, Stephen Hanson, Rick Herrmann, Ted Hopf, Andrew Lawrence, Jack Levy, Mark Lichbach, Brian Pollins, Bert Rockman, Janice Stein, and Steve Weber.

manner that, though inadequate, is not counterproductive to the ends it seeks. More fundamentally, by endorsing an arbitrary or inadequately theorized telos, it may sponsor a project that by its very nature is unrealistic.

2.2   King, Keohane, and Verba

Our volume is not conceived of as a critique of *Designing Social Inquiry,* but all our authors play off of it, and many use their criticisms as the jumping off point for their own arguments. KKV is the obvious foil because it is the most widely used text in graduate courses in method. It exudes a neopositivist confidence, shared by the many mainstream social scientists, that evidence is relatively unproblematic and can be decisive in resolving theoretical controversies. It emphasizes the existence of a single scientific method, the search for regularities, the issue of replication, the primacy of causal inference, the importance of 'observable' implications that are impartial to competing theories, and the significance of falsifiable hypotheses that are neutral between warring value commitments. It is regarded by its advocates as an important rejoinder to interpretivists, culturalists who flirt with postmodern relativisms, structuralists who have or have not found a haven in the now-dominant realist philosophy of science, and even rationalists (e.g., Hausman in the philosophy of economics literature) who have expressed doubts about the evidentiary basis of economics.[3]

KKV is also an easy target. It makes what many see as unwarranted claims for the rigor and success of quantitative research in the social sciences, unfairly deprecates qualitative research, and insists that qualitative researchers have much to learn from their quantitative colleagues.[4] Still others feel uncomfortable about the way in which KKV represent their protocols as hard-and-fast rules when, as is often the case, they are violated for good reason. A case in point is their injunction against selecting on the dependent variable. In his chapter, David Waldner provides a stunning example of how this strategy has been used successfully. Critics of neopositivism—including some of our contributors—contend that KKV misrepresents philosophical debates concerning falsification and science; it also fails to recognize that science is a practice based on conventions, not deductively established warrants, and that prediction is only one form of knowledge.

KKV is the appropriate starting point for this introduction. By describing what our contributors find valuable and objectionable in the book, we can compare their positions on important questions of method, epistemology, and ontology. When we do this, an interesting pattern emerges. Those closest to KKV in their orientation are equally keen to disassociate themselves from its epistemology and ontology. They

---

[3]Hausman, *Inexact and Separate Science of Economics.*

[4]Review Symposium: The Qualitative-Quantitative Disputation: Gary King, Robert O. Keohane, and Sidney Verbas's *Designing Social Inquiry*; Brady and Collier, *Rethinking Social Inquiry.*

do so to salvage methods and procedures they think valuable, but also to broaden the methodological menu and to confront problems with statistical inference to which KKV are oblivious. These contributors—Pollins, Waldner, and to a lesser extent, Chernoff—advocate an understanding of science that shows remarkable similarities to that advanced by more radical critics of KKV's project.

King, Keohane, and Verba explicitly acknowledge the importance of solid philosophical foundations. This makes it all the more surprising that they anchor their project in a version of logical positivism developed by the so-called Vienna Circle, a version that has long since been rejected by some of its key formulators and philosophers of science. Their choice is indefensible, but perhaps explicable in light of their belief in the unity of sciences and its corollary that the goals and methods of inquiry into the physical and social worlds are fundamentally the same. It is therefore appropriate to begin with a discussion of foundational claims and the reasons why the search for them is bound to fail.

2.3   Foundational Claims

Logical positivism was an attempt to provide a logical foundation for science. Its early propagators included Moritz Schlick, Otto Neurath, Rudolph Carnap, Herbert Feigl, and Kurt Godel. They assumed a unity among the sciences, physical and social, and sought to provide warrants for establishing knowledge. Toward this end, they established the "verification principle," which held that statements of fact had to be analytic (formally true or false in a mathematical sense) or empirically testable. It was soon supplanted by the principle of 'falsification' when Karl Popper, a close associate of the Circle, demonstrated that verification suffered from Hume's "Problem of Induction." For Popper, a scientific theory had to be formulated in a way that made it subject to refutation by empirical evidence. Scientists had to resist the temptation to save theories by the addition of ad hoc hypotheses that made them compatible with otherwise disconfirming observations. By this means, Popper asserted, a theory that was initially genuinely scientific—he had Marxism in mind—could degenerate into pseudoscientific dogma.

The Vienna Circle and Karl Popper had relatively little influence on the hard sciences but provided the ideological underpinning of the so- called behavioral revolution, of the 1960s. As Brian Pollins notes, their influence grew among social scientists, just as their ideas came under serious challenge by philosophers of science. One important reason for this challenge was the logical distinction that 'falsificationism' made between theory and observation. Carl Hempel demonstrated that no such distinction exists; tests cannot be independent of theory because all observations presuppose and depend on categories derived from theory. Unity of science was also questioned as the several sciences confronted different degrees of contingency in their subject matter. They worked out diverse sets of practices to deal with this and other problems and to collect and evaluate evidence. As Bernstein et al. point out, thoughtful social scientists, among them Max Weber, had come to recognize that regularities in human behavior and the physical world are fundamentally different. Social scientists, Weber argued, have a short half-life because they disappear or change as human goals and strategies evolve, in part because people come to understand these regularities and take them into account in their deliberations and strategies.[5] By the 1950s, Popper had come to understand "covering laws" as limited in scope, and perhaps as even unrealistic.[6] If he were alive today, he might well agree with Pollins that the social sciences are "the *really hard* sciences."[7]

KKV claim that 'falsifiability' lies at the heart of the scientific project and insist that they draw their understanding of it from Popper's 1935 book, *The Logic of Scientific Discovery*. This is the version, Pollins reminds us, that Popper later disavowed when he realized the problematic nature of evidence. For the same reasons, it calls KKV's project into question; at the very least it demands a thoroughgoing reformulation. The logical positivism on which KKV draws assumes a "real world" (i.e., an objective reality) that yields the same evidence even to investigators who search for it in the proscribed manner. This world is also expected to yield 'warrants' that validate theories on the basis of evidence and statistical tests. Knowledge is accordingly a function of good research design and good data.

The notion of a "real world" is very difficult to defend; and among our contributors, only Fred Chernoff makes the cases for a limited kind of 'naturalism.' Without a "real world," warrants for knowledge cannot be deduced logically, and efforts by philosophers to establish foundational claims, by either substantive (metaphysical) or epistemological (Kantian) means, must, of necessity, end in failure. If "unity of science" is indefensible, there are no universal procedures for determining what constitutes evidence or how it is to be collected and evaluated.

Alfred Schutz observed that all facts are created by cognitive processes.[8] John Searle distinguished between 'brute,' or observable facts (e.g., a mountain), and 'social,' or intentional and institutional facts (e.g., a balance of power).[9] Every social scientist deals primarily in social facts and must accordingly import meaning to identify and organize evidence. This is just as true of statistical evidence as it is of case studies. James Coleman has shown that every measurement procedure that assigns a numerical value to a phenomenon has to be preceded by a qualitative comparison. While the assignment of numbers may permit powerful mathematical

---

[5]Weber, "'Objectivity' in Social Science and Social Policy."

[6]Covering laws describe a model of explanation in which an event is explained by reference to another through an appeal to laws or general propositions correlating events of the type to be explained *(explananda)* with events of the type cited as its causes or conditions *(explanantia)*. It was developed by Carl Hempel in 1942 and derives from Hume's doctrine that, when two events are said to be causally related, all that is meant is that they instantiate certain regularities of succession that have been repeatedly observed to hold between such events in the past.

[7]Pollins, "Beyond Logical Positivism: Reframing King, Keohane, and Verba's *Designing Social Inquiry.*" pp. xx.

[8]Schutz, "Common-Sense and Scientific Interpretation of Human Action," p. 5.

[9]Searle, *Construction of Social Reality*.

transformations, it is illicit to make such assignments if the antecedent qualitative comparison has not or cannot be completed.[10] Many mainstream social scientists who acknowledge this problem nevertheless contend that even when the preconditions for successful measurements or causal modeling are not present, the "scientific method" should still serve as a regulative idea. Such a statement has no obvious meaning.

The foundational claims of logical positivism have been used by social scientists to serve political as well as intellectual ends. In the 1950s and 1960s, they were used to justify the behavioral revolution and its claims for institutional dominance and funding. Today, they defend orthodoxy against challenge while obscuring relations of power. Science and pluralism—and the former is impossible without the latter—demand that they be jettisoned.

What are we to do in the absence of a real world, unity of science, and foundational claims that could supply warrants? Does anything go, as some postmodernists joyously proclaim and some mainstream social scientists lament? None of our contributors believe that the baby of science has to be thrown with the bathwater of positivism. They advocate an understanding of science that has become widespread among philosophers and scientists: science as a set of shared practices within a professionally trained community.[11] Those sciences diverge in many ways, including in their relative concern for historical explanation versus prediction. Geology, pathology, and evolutionary biology are focused on the historical explanation of how the earth, dead people, and species came to be the way they are. Physics and chemistry use prediction as the gold standard and, unlike the sciences noted above, understand explanation and prediction to be opposite sides of the same coin.

The competent speaker, not the grammarian, is the model scientist, and each practitioner of discipline, like each speaker of a language, is the arbiter of its own practice. All insights and practices, no matter how well established, are to be considered provisional and almost certain to be superseded. Debates are expected to scrutinize tests and warrants as much as research designs and data. Consensus, not demonstration, determines what theories and propositions have standing. In his last decades, Popper came around to this position. He spoke of relative working truths— "situational certainty" was the term he coined—and emphasized the critical role of debate and radical dissent among scientists.[12]

Kratochwil suggests, and Pollins concurs, that the court is an appropriate metaphor for science as practice. As in court, difficult questions must be decided on the basis of evidence and rebuttal, not on the basis of proofs. Such contests are also quasi-judicial because they are subject to constraints that govern the nature of information and tests that can be presented to the jury. Those scientists who play

---

[10]Coleman, *Introduction to Mathematical Sociology.*

[11]Kuhn, *Structure of Scientific Revolutions;* Rouse, *Knowledge and Power;* Kratochwil, "Regimes, Interpretation, and the 'Science' of Politics."

[12]Popper, *Objective Knowledge,* pp. 78–81.

formal roles in such proceedings (e.g., journal editors, conference chairs), are, like judges, expected to adhere to well-established procedures such as blind peer review to promote fairness and to avoid conflicts of interest. Courts allow appeals that can be made on the basis of new evidence or improper treatment of the existing evidence or the disputing claimants. Science does the same and, in addition, also allows claims to be reopened on the basis of new insights concerning causal mechanisms. David Waldner provides a striking example of how this worked in the case of plate tectonics. The theory of continental drift was proposed by Alfred Wegener in the 1930s, but it was rejected by the scientific community because it ran counter to the prevailing orthodoxy that the continents were fixed. Wegener also hurt his case by failing to offer any plausible mechanism to explain continental drift. The debate was reopened in the 1960s, partially as a result additional evidence, but primarily in response to the appearance of a credible causal mechanism: thermodynamic processes deep within the earth that create convection currents that move the plates on which the continents rest.

Scientists recognize that the ethics of practice is at least as important as the logic of inquiry. Individual scientists must exercise care and honesty in developing frameworks and in collecting, coding, and evaluating data and communicating results to other members of the community. They must be explicit about the normative concerns and financial interests, if any, that motivate their work. Those who control funds, publications, appointments, tenure, promotions, honors, and the like must be open to diverse approaches, supportive of the best work in any research tradition, and committed to the full and open exchange of ideas. In the words of Rom Harre, science is "a cluster of material and cognitive practices, carried on within a distinctive moral order, whose characteristic is the trust that obtains among its members and should obtain between that community and the larger lay community with which it is interdependent."[13]

2.4 The Product of Inquiry

A common understanding of the nature of science does not necessarily promote a shared understanding of what is possible to discover. The hypothetical-deductive (H-D) method and mainstream social science in general assume that a self-correcting process of conjectures and refutations will lead us to the truth. Fred Chernoff, who is the most sympathetic among our authors to this understanding, argues that such a process will bring us closer to some truth. If progress is not possible, he asks, why would scholars continue to do research and engage in debate?

Brian Pollins recognizes that visions of the truth will always be multiple because different research communities will reach different conclusions about the nature of

---

[13]Harre, *Varieties of Realism*, p. 6.

knowledge, how it is established, and how it is presented. He is nevertheless convinced that adherence to the principles of falsifiability and reproducibility could foster more meaningful communication across these traditions and improve their respective "tool kits" This would make truth claims more difficult to establish and easier to refute. Hopf shares this vision to a degree. He accepts Popper's notion of working truths and argues that both mainstream and interpretivist approaches could make more convincing, if still modest, truth claims if they engaged in extensive mutual borrowing. To deliver on its promises, the mainstream needs to adopt a more reflexivist epistemology. Interpretivists, who have the potential to deliver on their promises can do so only by incorporating many mainstream research methods.

Mark Lichbach offers a parallel vision. In his view, theory consists of research programs that invoke different causal mechanisms to build theories that describe lawful regularities. Evidence establishes the applicability of these models of a theory for the models of data that exist in particular domains; the elaboration of a theory thus delimits the theory's scope. Evaluation grapples with the problem that the science that results from following the first two principles is prone to nonfalsifiability and to self-serving confirmations. Confrontations between theory and evidence are thus evaluated in the context of larger structures of knowledge, so rationalist, culturalist, and structuralist approaches in practice forge ahead on their own terms.

Kratochwil adopts a more radical position. If truth is no longer a predicate of the world—that is, not out there waiting to be discovered—then neither the H-D nor any other kind of research method can discover it. Truth is a misleading telos. We must rethink our goals and metaphors. Positivists conceived of truth as a chain that justifies beliefs by other beliefs, which ultimately must be anchored in some foundation. The mainstream, and some of our contributors, envisages truth to be more like a circle, whose area can be estimated with increasingly greater accuracy by approximating its circumference by use of successive polygons. This metaphor, Kratochwil suggests, is inappropriate because a circle is bounded by a perimeter, while the physical and social worlds have no knowable limits. If we need a metaphor, the game of Scrabble may be a more useful one. We begin with concepts and rules that make many outcomes possible. We can criss-cross or add letters to existing combinations, but all these entries must be supportive and must at least partially build on existing words and the concepts that underlie them. When we are stymied, we must play elsewhere but might by a circuitous route link up with all other structures. A modified game of Scrabble in which the board had no boundaries and new words could be placed anywhere might capture the idea even more effectively. According to this metaphor—in its original or modified form—progress in the social sciences is measured in terms of questions, not answers.

Bernstein, Lebow, Stein, and Weber share Kratochwil's ontology. They contend that all social theories are indeterminate because of the open nature of the social world. They offer an analogy between social science and evolutionary biology. Outside of certain "red states," evolution is widely regarded as a wonderfully robust scientific theory. Yet, it makes few predictions because its adherents recognize that almost everything that shapes the biological future is outside of the theory. It is the result of such things as random mutations and matings, continental draft, changes in the earth's precession and orbit, variations in the output of the sun—and how they interact in complex, nonlinear ways. Evolution is the quintessential example of a process where small changes can lead to very large divergences over time. The late Stephen Jay Gould suggested that if the tape of evolution could be rewound and played again and again, no two runs would come out the same.[14] Bernstein and his coauthors contend that this is also true of international relations, where personality, accidents, confluence and nonlinear interactions—all of which are, by definition, outside any theory of international relations—have a decisive influence on the course of events. Predictive theory is impossible, and so are even probabilistic theories—if they were possible, they would tell us nothing about single cases.[15]

Bernstein et al. recognize that human beings at every level of social interaction must nevertheless make important decisions about the future. They make the case for forward 'tracking' of international relations on the basis of local and general knowledge as a constructive response to the problems they, and other authors in this volume, identify in backwardlooking attempts to build deductive, nomothetic theory. They regard this kind of scenario construction, evaluation, and updating as a first step toward the possible restructuring of social science as a set of case-based diagnostic tools.

None of our contributors rally in support of KKV, but Chemoff offers a limited defense for the unity of science, contending that many of the methods used in the physical sciences are applicable to the social world. Despite the many problems involved in bridging the physical and social worlds, outright rejection of unity of science, he warns, involves even greater logical and methodological difficulties. To circumvent the problem of foundational claims, he draws on the understanding of the truth developed by American pragmatists. Following James, he suggests that to describe a statement is true is nothing more than saying that "it works" The concept of something working is treated at length by Peirce and James, and defined as something that helps us navigate the sensible world. This is not a correspondence theory because facts for James are nothing more than mental constructs that are maintained because of their demonstrable utility. In his understanding, there is no useful belief that does not accord with the 'facts.' Even traditional correspondence theories, Chernoff suggests, frame truth as a relationship between a statement and external reality, as opposed to a feature of reality itself. They are accordingly testable against our observations, as these observations in turn constitute the 'effects's of reality. Unlike Platonism, which views the truth as a form, correspondence theories, Chernoff insists, are not vulnerable to Kratochwil's argument that truth is not a predicate of the world.

The previous discussion makes clear the division among our contributors concerning the nature of knowledge. Some, such as Pollins and Chernoff, believe that good questions, methods, and evidence can lead us to some kind of knowledge.

---

[14]Gould, *Wonderful Life.*

[15]For a thoughtful rebuttal of this argument, see Waldner, 'Anti-Determinism.'

Others, such as Kratochwil and this editor, believe that all but the most banal propositions can ultimately be falsified, but the process of falsification requires us to develop new research tools and questions. Falsification can lead us to more sophisticated propositions and methods.[16]

2.5 The Purpose of Inquiry

Mainstream social science envisages the goal of inquiry as knowledge, and many of its proponents believe that knowledge requires fact to be separated from values. "Value neutrality" is often described as one of the attributes of true science. It follows that research questions should grow out of prior research or empirical discoveries. The 'fact-value' distinction dates back to David Hume, who insisted that statements of fact can never be derived from statements of value, and vice versa. His argument and its implications have been debated ever since. They were a central feature of the *Methodenstreit* that began in Vienna in the late nineteenth century.

Max Weber, one of its most distinguished participants, made the case for the social sciences being fundamentally different from their natural counterparts. Values neither could nor should be separated from social inquiry. This would represent an attitude of moral indifference, which he insisted, "has no connection with scientific 'objectivity'"[17]

All of our contributors side with Weber on the fact-value distinction. Jack Levy and Andrew Lawrence, who hold quite different views about the value of the democratic peace research program, agree that its ultimate justification must be the insights and guidance it offers us about reducing the frequency of violent conflict. It is possible to emphasize either facts or values in research, but problems arise when either is pursued at the expense of the other. Value neutrality is impossible for there is no way we can divorce our normative assumptions and commitments from our research, and attempts to do so are damaging to discipline and society alike. Efforts to segregate research from values have ironically encouraged and allowed scholars to smuggle norms into their research through the back door. According to John Gerring, the adoption of a Pareto optimality, is a case in point. It is not a scientific choice but a partisan and highly consequential moral choice.[18]

Normative theorizing must deal with facts just as empirical research must address norms. They do no inhabit separate worlds. Nor should they, because the purpose of social science is practical knowledge. The choice of subjects and methods presume judgments of moral importance. It is incumbent upon researchers

---

[16]Maher, *Betting on Theories*, p. 218, makes the same assertion about the sciences, whose history, he claims, "is a history of false theories."

[17]Weber, *Methodology of the Social Sciences*, p. 60.

[18]Gerring, "A Normative Turn in Political Science?"

to make their values or telos explicit and fair game for analysis and critique. In the broadest sense, political science can be described as the application of reason to politics. It is practiced by people with the requisite expertise, which includes the ability to separate reason from values in their analysis—although not in their choice of topics. Hume's 'fact-value' distinction can be distorted at either extreme: either by denying values or by denying facts. We need to maintain the distinction but bring norms into the foreground, not only in research, but in our training of graduate students.

A more serious problem arises from the failure of Hume's dichotomy to capture what John Searle has called "institutional facts." These are neither facts nor values, but 'performatives'—like the "I do" of a marriage ceremony—that establish actors and their relationships. It is not far-fetched to argue that the most interesting questions of the social and political world are 'outside' the Humean dichotomy, and that social science must also go beyond it. Weber, for one, recognized that values are not just the preferences of researchers but are also constitutive of their identities and interests. For John Searle, they are the glue that holds society and its projects together. If we want to understand society, we need to adopt methods that confront values and their importance, not rule them out a priori as much of mainstream has tried to do.[19]

In large part, differences over the role of values reflect differences in the purposes of inquiry. Neopositivists who envisage theory as an end product of social science sometimes see values as a distraction and embarrassment. They would believe, like physical scientists, that their research is driven by puzzles and anomalies that arise from their research. This ignores the well-documented extent to which research agendas of physical scientists are equally driven by normative commitments. More thoughtful neopositivists, including the contributors to this volume, see nothing wrong with acknowledging the normative and subjective nature of research agendas. What makes their research scientific is not their motives but the rigor of their methods. Further along the spectrum are nonpositivists, at least some of whom regard theory as a means to an end and as valuable only in so far as it helps us understand and work through contemporary political, economic, and social problems. For them, social science begins and ends with values.

2.6   The Method of Inquiry

Contributors who are generally sympathetic to the goals of mainstream—Pollins, Chernoff, Waldner, and Levy—consider KKV's depiction of research as a misguided attempt to put the scientific method into a statistical straitjacket. KKV equate good research design with inference and define it in a way that makes it all but synonymous with statistical inference.

---

[19]Searle, *Construction of Social Reality*, pp. 27–28.

For KKV and others who subscribe to their narrow framing of the Hypothetical-Deductive (H-D) method, the only ways to challenge a theory are by disputing its internal logic or by adding additional observations. Kratochwil, Hopf, and Waldner all recognize that adding observations addresses the first problem of induction raised by Hume: "How much is enough?" It says nothing about the second problem: causality. The discovery of laws requires leaps of imagination; laws are not simply statements of regularities, but creative formulations that order those regularities or make their discovery possible. Both theory formation and testing frequently require and certainly benefit from the use of counterfactual thought experiments.[20]

The core principle of mainstream social science is the H-D model. KKV's good scientist "uses theory to generate observable implications, then systematically applies publicly known procedures to infer from evidence whether what the theory implied is correct."[21] Valid observations are all that is required to test a theory, and a single, critical experiment can refute a law. In practice, David Waldner observes, a variety of criteria are used to confirm and disconfirm theories, of which evidence is only one. This is evident from the solution of the mystery of dinosaur extinction, the very example that KKV improperly cite as an outstanding success of the H-D method. They claim that the hypothesis of a meteor impact led to the search for iridium, whose discovery at the K/T boundary confirmed the hypothesis. In fact, researchers reasoned backwards, from the discovery of the iridium layer to its probable cause, and focused on causal mechanisms—what it would take to kill dinosaurs and produce iridium—rather than on research design considerations. Meteor impact is now generally accepted by the wider scientific community—because of the causal mechanism and logic that connects it to an otherwise anomalous outcome. Dinosaur extinction is also an interesting case because it violates KKV's supreme injunction against coding on the dependent variable. Walter Alvarez and the Berkeley group did just this; they never examined other instances of mass extinction and failed to study epochs of non-extinction when extraterrestrial impacts were common. They also ignored far more numerous sub-extinctions.

Drawing on work in analytical philosophy, Waldner distinguishes between inferences and explanations. He suggests that we evaluate hypotheses in terms of their evidentiary support and theoretical logics. A confirmed hypothesis is one that has survived scrutiny against its closest rivals—given the current state of theory and evidence. It is more reasonable than disbelief but still subject to revision or refutation. We explain by using confirmed hypotheses to answer questions about why or how phenomena occur. All explanations require confirmed inferences, but not all inferences constitute explanations or embody them. Causal mechanisms can impeach or enhance hypotheses with otherwise impeccable research- design credentials. They promote inferential goodness via theory, not via research design.

---

[20]Weber, "Counterfactuals, Past and Future"; Lebow, "What's So Different about a Counterfactual?"

[21]King, Keohane, and Verba, "The Importance of Research Designs in Political Science," p. 476.

Waldner offers seven ways in which causal mechanisms can be used to reject hypotheses. His major point is that there are many ways to confirm and reject hypotheses, only one of which is statistical inference. He agrees with Hopf that underdetermination is not resolved by collecting more evidence, but by better understanding the evidence we already have. Good social science seeks contextualized explanations based on causal mechanisms, not just law-like regularities.

Theories are also rejected because better theories come along. The Ptolemaic model of planetary motion successfully accounted for the motions of the sun, moon, and five known planets. It was rejected in favor of Copernicus's heliocentric model because the latter was simpler; Ptolemy's model required eighty epicycles to explain these motions. His system was nevertheless *more* accurate than that of Copernicus and remained so until Kepler's Laws could augment the latter.

In practice, most refutations are not accepted, but understood as problems of measurement, experimental error, "put right" through manipulation of data or explained away as anomalies. The hole in the ozone layer over the south pole offers a nice example. The British Antarctic Survey began taking measurements of the density of the ozone layer in 1957, and—for the first twenty years—variation followed a regular seasonal pattern. Beginning in 1977, deviation from this pattern was noted, and at first attributed to instrument error. Every spring, the layer was measured as weaker than the previous spring, and by 1984, scientists reluctantly concluded that change was occurring. This conclusion met considerable resistance until experiments and observations revealed that industrial chemicals, particularly chlorofluorocarbons CFCs containing chlorine, could destroy ozone. Refutations are taken seriously only when reasons are provided for why the observed deviations were systematic and not due to random errors or disturbances, and ozone depletion was no exception. Even then, as research on deterrence indicates, refutations can encounter serious resistance when the theories in question serve important political or psychological ends.[22]

There may be good reasons for ignoring refutations. Paul Diesing reminds us that every theory is refuted, as they all are at least somewhat false. If we give up theories because they are refuted, we can no longer profit from their heuristic potential to produce better theories.[23] It may be, as Imre Lakatos suggests, that occasional, if partial, verifications of theories are what keep research programs going, and they are all the more necessary when their theories have been exposed by repeated refutations.[24]

Pollins, Hopf, Waldner, and Chernoff, all offer suggestions for overcoming methodological and epistemological narrowness. Pollins insists that there is no logical reason why the rules of scholarship cannot be pluralistic. Many of the practices described by KKV can be incorporated into a "new and broader based social science epistemology." For Pollins, the two defining criteria of such a science

---

[22]Lebow and Stein, *We All Lost the Cold War,* chs. 4 and 13; Kull, *Minds at War.*

[23]Diesing, *How Does Social Science Work?* p. 45.

[24]Lakatos, "Falsification and the Methodology of Scientific Research Programmes," p. 137.

are falsifiability and reproducibility. Falsifiability assumes that we do the best we can to be clear, and that "more correct" can be distinguished from "less correct." Observations should be classified as consistent or inconsistent with a claim, and decisive tests ruled out because of the theory-laden nature of observation. Falsifiability is a communicative concept that allows challenges to and changes in conceptual categories. So is reproducibility. It requires research to be described in ways that allows duplication so others can try to obtain the same results from the same evidence, or same kind of evidence. Such an approach, Pollins acknowledges, shifts the emphasis from the interaction between theory and observation to that between claimant and professional audience.

Hopf plays variations on this theme. He stresses how much the mainstream and interpretivist traditions actually share, and he identifies seven key methodological conventions in this regard: differentiate premises from conclusions and correlations from causes, respect the canons of inference, establish standards of validation for data and other source materials, address problems of spuriousness that arise from correlations, rely on syllogistic and deductive logic, and accept the contestability of all beliefs and findings. Hopf suggests that differences within the reflexivist community on these issues are more serious than those between it and the mainstream. The deepest cleavage runs between phenomenological, interpretevist, and hermeneutic approaches on the one hand, and some postmodern or critical approaches on the other. Some representatives of the latter maintain that narration constitutes its own truth and has no need of argument or proof. So-called mainstream reflexivists are interested above all in the ways in which social order reproduces itself through the behavior of actors. To do so they must consider the context and meaning in which these interactions take place and the various ways in which observers can come to understand them. They have the same need as mainstream scholars to consider the nature of facts, evidence, truth, and theory.

2.7   The Practice of Inquiry

Science consists of hard-fought bull sessions with students and colleagues, applications for funding, the conduct of research, management of research facilities and teams, writing up research results, and the presentation of findings. Findings may be circulated as draft papers, posted on the Web as preprints, or submitted to journals or publishers as would-be articles or books. Contested claims are adjudicated at many of these steps by researchers themselves, in informal discussions among colleagues, the more the formal proceedings associated with peer review, panel presentations, and debates on Web sites and in professional publications. Such a process is quite distinct from rarefied debates—such as those in this volume—about the nature and purpose of inquiry and the methods appropriate to it.

To understand the practice of science, we need to adopt a microperspective; and with that end in mind, we asked two contributors to look into why some research programs are successful. Jack Levy—guilty of coding on the dependent variable here—examines three successful paradigms in international relations: rational choice, territory-war and power balance, and *democratic peace* (DP). He evaluates KKV's contention that there is little tension between normative and descriptive research programs, and that the most successful programs are those with the most empirical support. Andrew Lawrence devotes his chapter entirely to the democratic peace.

Levy finds that research programs in international relations are sustained by different combinations of incentives. Rational choice is largely theory driven, while territory-war and the power balance, and DP are more evidence driven. Research programs propelled by a powerful, or at least, intellectually appealing, theory can become self-sustaining even in the absence of evidence. This is true of general equilibrium theory in economics and rational choice in political science, although the latter's influence expanded considerably when it was linked to the quantitative research tradition and received, in Levy's judgment, considerable empirical confirmation. All three research programs indicate that any assessment of the relative importance of theory and evidence in sustaining a research program will depend on the level of theory at which we focus. A paradigm may be theory driven (e.g., liberalism), but a theory within it may be evidence driven (e.g., DP). Research programs can also be motivated in sequence by theory and evidence.

Scholarship in the war/territory and DP research programs has responded to the discovery of striking empirical regularities: that a disproportionately high number of wars involve territorial disputes, that territorial disputes are more likely to lead to wars than to any other kind of dispute, and that democracies appear never to go to war with one another. Levy acknowledges that normative concerns have also influenced the prominence and evolution of the DP research program. He nevertheless questions what he describes as the widely held view that policy agendas account for the appeal and popularity of DP. Although many liberals were drawn into the research, they have not allowed their values or political preferences to stand in the way of or distort the evidence. One reason for this is the engagement in this research of scholars from other political perspectives. In Levy's judgment, the DP is a quintessential example of a 'progressive' and responsive research program.

Lawrence is critical of the DP program. In his view, it has distorted its Kantian origins and yielded diminishing returns theoretically. The prevailing norms of mainstream social science—especially those of quantitative social science—have restricted the debate, led to a fetish with numbers and acceptance of "common sense" definitions of key variables such as democ and war. These definitions obscure the meaning of these variables and how these meanings have evolved over time. Statistical tests are largely use because "generous fudge factors" are used to code borderline cases of both war and democracy. Quantitative researchers on the whole emphasize external validity (comparison across cases) over internal validity—the application, or fit, of measures to individual cases. Causal inference is sup to permit communication across the discipline, but the DP research program narrows it. In Lakatosian definition it is a "degenerate paradigm."

Lawrence contends that unarticulated but critical normative presuppo and commitments often drive research. He is struck by the political bias of the DP literature, the enthusiasm the research program has generated among liberals, and the claims by some that the DP is one of the most robust research findings in international relations. In his view, its focus on nonwar among democracies, conceptions of democracy, and codings of war and democracy reflect, at best, parochial, and at worst, self-serving, perspectives that make the research program a justification for America's foreign policy and way of life. Democratic peace researchers have framed their inquiry in a way that excludes cases where the United States has resorted to force. The program reflects the general tendency of American social scientists to employ positivism as a means of evad reflexive self-knowledge. In this sense, DP, like deterrence theory during the cold war, is best understood as part of the phenomena these theories seek to explain.

2.8   'Social' Knowledge

Social understanding is inherently subjective. Research agendas, theories, and methods are conditioned by culture, beliefs, and life experiences. So too is receptivity to research findings. Recognition of this truth has led some postmodernists to interpret science as a political process and cloak for individual and group claims to privilege. This view of science is one because it ignores the barriers erected by the scientific method against theories and propositions that either cannot be falsified or are demonstrably false.

The scientific method does not always prevail over politics and preju. The problem is sometimes the scientists themselves. Nineteenth-century biological and anthropological studies of cranial capacity 'proved' the superiority of the Caucasian 'race.' Some contemporary researchers are still trying to do this with data from intelligence tests. Well-founded scien claims also encounter resistance from the wider community. The of evolution continues to provoke widespread opposition from fundamentalist Christians. Claims by medical researchers that smoking is harmful, and more recently, by environmental scientists that the waste products of industrial society threaten to produce an irreversible transform of the environment, have encountered predictable opposition from industries with profits at stake. The tobacco companies and some major polluters support scientists who dispute these claims.

Our contributors make it clear that there is no such thing as a "scien method." Researchers and philosophers of science argue over what constitutes adequate specification and testing, the extent to which it is pos, and, more fundamentally, about the nature and goals of science. Attempts to provide definitive answers to these questions, as Karl Popper recognized, inevitably fail and risk substituting dogma for the ongoing questioning, inquiry, and debate that constitute the core commitment of science. These controversies render scientific truth uncertain, but working scientists, invoking the techniques and skills they have learned, generally have little difficulty in distinguishing good from bad science.

   The scientific method in many ways resembles the Bill of Rights of the American Constitution. Its meaning is also interpreted through practice. And, like the scientific method, it has not always been interpreted or applied fairly. The Bill of Rights has sometimes failed to protect political, religious, and so-called racial minorities from the ravages of prejudice. In 1898, *Plessy* versus *Ferguson* established the principle of separate and equal education for African Americans that endured until *Brown* versus *Board of Education* in 1954. De facto segregated education continues to this day in some locales. *Brown* versus *Board of Education* reflected changing attitudes toward African Americans and the constitution itself. Another impetus was extensive social science research that demonstrated that separate edu was inherently unequal. Despite continuous controversy about the meaning of the constitution and despite periodic failures to apply its prin in practice, there is an overwhelming consensus that the Bill of Rights—and even more importantly, the American public's commitment to tolerance—remains the most important guarantee of individual free. The scientific method is an imperfect but essential bulwark against many of the same kinds of passions. Like the Constitution, it ultimately depends on the ethical standards and commitments of the community it serves.

   As many of our contributors have suggested, there is an important dis to be made between the questions we ask and the ways in which we answer them. What distinguishes us from ideologues is our commit to finding and evaluating answers by means of a scientific method. Social scientific research agendas are shaped by political beliefs, life experi, and desires for professional recognition. There is nothing wrong with these motives. Good social science should be motivated by deep per involvement in the burning issues of the day. Research can clarify these issues, put new issues on the agenda, and propose and evaluate the consequences of different responses. It can also influence the way people conceive of themselves, frame problems, and relate to the social order.

   Logical-neopositivism and other "unity of science" approaches risk making social science sterile in its search for passionless, abstract truths. Some forms of postmodernism would make social science irrelevant by its rejection of the scientific method and insistence that all 'readings' of texts and the world at large have equal standing. Social scientists need to confront both these dangers by reaffirming and explaining their twin commitments to social progress and the scientific method. The links between ourselves and our research do not undercut our claim to be practicing science, they make us better scientists and human beings.

2.9   The Structure of the Book

Our volume contains eleven chapters divided into four sections. This introduction is followed by chapters by Fritz Kratochwil and Ted Hopf on foundational claims. They develop ontological and epistemological criti of the unity of science. Kratochwil shows why warrants can neither be taken for granted nor derived from theories of science. Hopf argues that social science must become reflexivist in its epistemology.

he next section, on the product of inquiry, consists of chapters by Brian Pollins and Fred Chernoff. Pollins accepts the gist of the Kratochwil-Hopf criticisms and believes they point the way toward the possibility of a broader-based, pluralist epistemology that would permit and encourage diverse forms of research and knowledge building. Fred Chernoff reviews and assesses the epistemological and metaphysical claims of Kratochwil, Hopf, and Pollins as well as their take on natural. He makes a case for a pragmatic theory of knowledge and for a modified conventionalist account of social science as the best way of accounting for the successes and frequent failures of social science research during the past fifty years.

We then turn to the purpose and methods of research. David Waldner examines the role of causal logic in his explanation of science in general. He contends that they are the distinguishing characteristic of all explana, and are routinely used to enhance or undermine theories. The next two chapters attempt to evaluate ongoing and well-regarded research pro in light of earlier discussions. As noted earlier, Jack Levy analyzes three such programs, including the democratic peace, and provides a pos take on their accomplishments. Andrew Lawrence focuses on the democratic peace and finds it crippled by epistemological, methodologi, and normative problems.

In lieu of a conclusion, we offer two contrasting visions for the future of social science. Steven Bernstein, Ned Lebow, Janice Stein, and Steve Weber contend that our goal should be practical knowledge relevant to individual cases, as this would make our profession relevant to the policy world. They describe the benefits and procedures of scenario generation and updating as powerful forecasting tools that, at best, would provide useful guidance in addressing complex real-world problems and at least provide useful early warning of impending policy failure. Mark Lichbach draws on the arguments of this book and on observations gleaned from observing the practice of science to offer the outlines of an epistemology that would incorporate mainstream and interpretivist practices and encourage progress toward better theories in both traditions.

Bibliography

Coleman, James. 1964. *Introduction to Mathematical Sociology.* New York: Free Press.
Diesing, Paul. 1991. *How Does Social Science Work? Pittsburgh: University of Pittsburgh Press.*
Gerring, John; Joshua Yesnowitz. 2006. "A Normative Turn in Political Science?" *Polity* 38, 1, pp. 101-33.
Gould, Stephen J. 1990. *Wonderful Life: The Burgess Shale and the Nature of History.* New York: Norton.
Harré, Rom. 1987. *Varieties of Realism: A Rationale for the Sciences.* Oxford: Blackwell.
Hausman, Daniel M. 1992. *Inexact and Separate Science of Economics.* Cambridge: Cambridge University Press

Hempel, Carl. 1942. "The Function of General Laws in History." *Journal of Philosophy* 39, pp. 35-48.

King, Gary; Keohane, Robert O.; Verba, Sidney. 1994. *Designing Social Inquiry*; Brady and Collier, *Rethinking Social Inquiry*.

King, Gary; Keohane, Robert O.; Verba, Sidney. 1995. "The Importance of Research Designs in Political Science." *American Political Science Review* 89 pp. 454-81.

Kratochwil, Friedrich V. 1988. "Regimes, Interpretation, and the 'Science' of Politics." *Millennium* 17 (Summer): 263-84.

Kuhn, Thomas S. 1962. *Structure of Scientific Revolutions*. Chicago: University of Chicago Press.

Kull, Steven. 1988. *Minds at War*. New York: Basic Books.

Lakatos, Imre. 1970. "Falsification and the Methodology of Scientific Research Programmes," in Imre Lakatos and Alan Musgrave (eds.), *Criticism and the Growth of Knowledge*. Cambridge: Cambridge University Press, pp. 91-195.

Lebow, Richard Ned. 2000. "What's So Different About a Counterfactual?" *World Politics* 52 (July): 550-585.

Lebow, Richard Ned; Stein Gross, Janice. 1994. *We All Lost the Cold War*, chs. 4 and 13;

Maher, Patrick. 1993. *Betting On Theories*. Cambridge: Cambridge University Press.

Pollins, Brian. 2007. "Beyond Logical Positivism: Reframing King, Keohane, and Verba's *Designing Social Inquiry*," in Lebow and Lichbach, *Theory and Evidence*, pp. 87-106.

Popper, Karl. 1966. *Objective Knowledge*. Oxford: Oxford University Press, pp. 78-81.

Review Symposium: The Qualitative-Quantitative Disputation

Rouse, Joseph. 2005. *Knowledge and Power: Toward a Political Philosophy of Science*. Ithaca: Cornell University Press.

Schutz. 1954. "Common-Sense and Scientific Interpretation of Human Action." *Philosophy and Phenomenological Research*. 14: 1-38.

Searle, John R. Searle. 1995. *Construction of Social Reality*. New York: The Free Press.

Waldner, 2008. "Anti-Determinism: Or What Happens When Schrodinger's Cat and Lorenz's Butterfly Meet Laplace's Demon in the Study of Political and Economic Development," Paper presented at the annual meeting of the American Political Sciene Association, Boston, Ma. 2009

Weber, Max. 1968. *Gesammelte Aufsätze zur Wissenschaftslehre*, ed. Johannes Winckelmann, 3rd ed. Tübingen: J. C. B. Mohr (Paul Siebeck)

Weber, Max. 2012. in Hans Henrik Brunn and Sam Whimster, eds., *Max Weber: Collected Methodological Writings*. London: Routledge.

Weber, Steven. 1996. "Counterfactuals, Past and Future," in P.E. Tetlock and A. Belkin, eds., *Counterfactual Experiments in World Politics*. Princeton: Princeton University Press., pp. 272, 278.

Chapter 3
Social Science as Case-Based Diagnostics

Steven Bernstein, Richard Ned Lebow, Janice Gross Stein
and Steven Weber

3.1 Introduction

A deep irony is embedded in the history of the scientific study of political science, but especially of international relations.[1] Recent generations of scholars separated policy from theory to gain an intellectual distance from decision making to enhance the 'scientific' quality of their work. But five decades of well-funded efforts to develop theories of international relations have produced precious little in the way of useful, high confidence results. Theories abound, but few meet the most relaxed 'scientific' tests of validity. Even the most robust generalizations or laws we can state— war is more likely between neighboring states, weaker states are less likely to attack stronger states—are close to trivial, have important exceptions, and for the most part stand outside any consistent body of theory.

A generation ago, we might have excused our performance on the grounds that we were a young science still in the process of defining problems, developing analytical tools and collecting data. This excuse is neither credible nor sufficient; there is no reason to suppose that another fifty years of well-funded research would produce valid theory in the Popperian sense. We suggest that the nature, goals, and criteria for judging social science theory should be rethought, if theory is to be more helpful in understanding the real world.

---

[1]This text was first published as "Social Science as Case-Based Diagnostics," co-authored with Steven Bernstein, Janice Stein and Steven Weber, in, Richard Ned Lebow and Mark Lichbach, eds., *Political Knowledge and Social Inquiry* (New York: Palgrave, 2007), pp. 229–260. ISBN 9781403974563. The permission to republish this text was granted on 18 June 2015 by Claire Smith, Senior Rights Assistant, Nature Publishing Group & Palgrave Macmillan, London, UK.

© The Author(s) 2016
R.N. Lebow (ed.), *Richard Ned Lebow: Major Texts on Methods and Philosophy of Science*, Pioneers in Arts, Humanities, Science, Engineering, Practice 3,
DOI 10.1007/978-3-319-40027-3_3

We begin by justifying our pessimism, both conceptually and empirically, and argue that the quest for predictive theory rests on a mistaken analogy between physical and social phenomena. Evolutionary biology is a more productive analogy for social science.[2] We explore this analogy in its 'hard' and 'soft' versions and examine the implications of both for theory and research in international relations. We develop the case for forward 'tracking' of international relations on the basis of local and general knowledge as an alternative for backward-looking attempts to build deductive, nomothetic theory.

This chapter is not a broadside against 'modern' conceptions of social science. Rather, it is a plea for constructive humility in the current context of fascination with deductive logic, falsifiable hypothesis, and large-N statistical 'tests' of propositions. We propose a practical alternative for social scientists to pursue in addition, and in a complementary fashion, to 'scientific' theory-testing as traditionally conceived.

3.2 Overcoming Physics Envy

The conception of causality on which deductive-nomological models are based, in classical physics as well as in social science, requires empirical invariance under specified boundary conditions. The standard form of such a statement is this: given A, B, and C, if X then (not) Y.[3] This kind of bounded invariance can be found in closed, linear systems. Open systems can be influenced by external stimuli, and their structure and causal mechanisms evolve as a result. Rules that describe the functioning of an open system at time T do not necessarily do so at T + 1 or T + 2. The boundary conditions may have changed, rendering the statement irrelevant. Another axiomatic condition may have been added, and the outcome subject to multiple conjunctural causation. There is no way to know this a priori from the causal statement itself. Nor will complete knowledge (if it were possible) about the system at time T necessarily allow us to project its future course of developments.

In a practical sense, all social systems (and many physical and biological systems) are open. Empirical invariance does not exist in such systems, and seemingly probabilistic invariances may be causally unrelated (Bhaskar 1979; Harré/Secord 1973; Collier 1994; Patomaki 1996; Jervis 1997). As physicists are the first to admit, prediction in open systems, especially nonlinear ones, is difficult, and often impossible.

The risk in saying that social scientists can 'predict' the value of variables in past history is that the value of these variables are already known to us, and thus we are

---

[2]We use evolutionary biology as an analogy for modes of reasoning, not as a model of politics per se.

[3]We state the rule in this way to avoid the confusion of "affirming the consequent" (as in if X then Y) and thus to emphasize falsifiability.

not really making predictions. Rather, we are trying to convince each other of the logic that connects a statement of theory to an expectation about the value of a variable that derives from that theory.

As long as we can establish the parameters within which the theoretical statement is valid, which is a prerequisite of generating expectations in any case, this 'theory-testing' or 'evaluating' activity is not different in a logical sense when done in past or future time.[4]

Consider how this plays out in evolutionary biology, the quintessential open system. Evolution is the result of biological change and natural selection. The former is a function of random genetic mutation and mating. The latter depends on the nature and variety of ecological 'niches' and the competition for them. These are in turn shaped by such factors as continental drift, the varying output of the sun, changes in the earth's orbit, and local conditions that are hard to specify. Biologists recognize that all the primary causes of evolution are random, or if not, they! interact in complex, nonlinear ways and make prediction impossible. Certain kinds of outcomes can be "ruled out" in a probabilistic sense, but almost never absolutely. Biologists have attempted to document the course of evolution and explain the ways in which natural selection works. Historical and theoretical work has resulted in a robust theory of evolution; that permits scientific reconstruction of the past in the context of a logic that explains why things turned out the way they did.

One of the big controversies within this research community is about the contingency of that past. Stephen Jay Gould (1989) makes the case for determining the role of accident in evolution. He insists that if you could rewind the tape of life and run the program over again you would end up each time with a radically different set of organisms. Some of his colleagues find his claim extreme. Ever since Darwin, it has! been recognized that evolution produces morphological similitude because there is something like a 'best' set of physical characteristics and strategy for grappling with the challenges of life. Diverse species have converged independently on body plans and life styles that are suited to avoiding predators and to exploiting food resources.[5] What is at stake in this controversy is how close the system has come to optimality, and the extent to which factors outside the system (Gould 1989) or the system itself (Morris 1998) are most important in shaping the course of evolution. Both sides acknowledge that the primary causes of evolution are independent of and outside any theory of evolution.

The study of evolution has been approached from scientific and heuristic perspectives. The scientific approach should be of particular interest to political scientists because it eschews prediction in favor of explanation. Working on the assumption that the course of evolution is determined by chance and context, Charles Darwin and his successors developed a theory of process to understand the past. That theory and its extensions fully meet the accepted criteria of scientific theories; they consist of a set of linked propositions with well-specified terms and

---

[4]See Weber, "Counterfactuals Past and Future," in Tetlock and Belkin, eds, 1996.
[5]For elaboration, see Morris (1998).

domain and are thus empirically falsifiable. Darwinian theory, widely regarded as one of the seminal scientific advances of the modem era, challenges those political scientists who assert that prediction is the principal, or even only, goal and test of a scientific theory.

The heuristic approach to evolution consists of narratives intended to influence our thinking about ourselves and our environment. These stories and the homilies associated with them have been extremely influential. What has sometimes been called the "Darwinian revolution" recast human conceptions of species 'uniqueness,' its relationship to other life forms, and hastened the trend toward secularization by providing an eminently plausible substitute for a deity-centered account of creation. More recent work on mitochondrial DNA, which suggests that Africa was the birthplace of *homo sapiens sapiens* and that 'Lucy' was our common ancestor, also have profound political and social implications that neither scientists nor journalists have been shy to draw. These examples stand in sharp contrast to the nineteenth-century use of evolution to justify war and imperialism and prop up Western claims of racial superiority. Gould (1996) has shown how many textbook treatments of evolution are still "species centric" and contain illustrations that show humanity as the apex of evolutionary development.

There is a nice correspondence between the heuristic forms of evolutionary biology and international relations. Narratives of international relations also encapsulate so-called lessons of the past—the more recent past, to be sure—to influence thinking about the present and future. Like homilies about evolution, scholars, journalists, and policymakers cite history as a general guide to action (e.g., realism, deterrence, the dangers (or benefits) of armaments), or as justification for specific foreign policies. Proponents and opponents of intervention in Bosnia, Kosovo, and Iraq have attempted to legitimize their respective positions with reference to 1914, the failure of the League of Nations, the Holocaust, and Vietnam.

The scientific study of international relations fits best, if partially, with evolutionary biology. For fundamentally similar reasons, international relations theory will not be able to predict events, trends, or system transformations in a useful way. But international relations theory, like its Darwinian counterpart, can attempt—as many scholars do—to develop theories of process to organize our thinking about the past Like paleontologists reading the evidence of fossil beds, these scholars use documents and interviews with former policymakers to evaluate competing theories, qualitatively and quantitatively. Using theories as starting points, they can also reconstruct the origins of revolutions, wars, accommodations, and other international phenomena in cases where there is adequate contextual evidence about the goals, understandings, and calculations of relevant actors and the political environment in which they functioned. Explanatory theories that pass the same tests as evolution have a serious claim to scientific status. International relations differ in at least one major respect from biology A robust theory of evolution is I possible because the actors in this drama—plants, animals, and other forms of life—know nothing about the theory Human beings devote enormous resources, individually and collectively, to understanding the nature of their environment That understanding has led them to interfere with biological evolution in important ways.

People started to domesticate and selectively breed animals at least 10,000 years ago. Intensive experimentation with crops started not long afterward In the twentieth century, we have utilized antibiotics and other medical techniques to interfere with natural selection, and knowledge of molecular biology to alter genetically a wide range of plants and animals. The current century will almost certainly bring more radical forms of bioengineering, including gene substitution and more general manipulation of the human genome.

Human intervention in the processes that govern social and political relations has been even more striking. As a general rule, the more people think that they understand the environment in which they operate, the more they attempt to manipulate it to their advantage. Such behavior can relatively quickly change the environment and its governing rules. The Asian financial crisis of the 1990s offers a good example. Rapid growth allowed some Asian countries to attract hundreds of billions of dollars of short-term international loans in the early 1990s. When short-term money managers began to lose faith in the Thai and South Korean economies, the IMF pressured their governments to maintain exchange rates by raising interest rates to restore investor confidence. Such a strategy had often worked in the past, yet the more the Asian government tried to defend their currencies, the more panic they incited. Money managers hastened to withdraw their funds before local currencies collapsed. Urged by the IMF and Washington, the Russian, South African, and Brazilian economies subsequently pursued the same policy with similar disastrous results. In the aftermath, the IMF and many prominent economists came to recognize that greater sophistication on the part of investors and the greater mobility of capital had changed the rules of the game. They needed different strategies to cope with the problem of investor confidence (Sachs 1998; Radelet/Sachs 1999).

Knowledge of structure and process also allows: conscious and far-reaching transformations of social systems. Smith, Malthus, and Marx described what they believed to be the inescapable! 'laws' that shaped human destiny. Their predictions were not fulfilled, at least in part, became their analyses of economics and population dynamics prompted state and corporate intervention designed to prevent their predictions from coming to pass. Human prophecies—which axe a form of prediction— are often self-negating.

A similar process has occurred in international relations. Prodded by two destructive world wars and the possibility of a third that might be fought with nuclear weapons, leaders sought ways to escape from some of the deadly consequences of international anarchy and the self-help systems it seemed to engender. They developed and nurtured supranational institutions, norms, and rules that mitigated anarchy and provided incentives for close cooperation among developed states. Gradually, the industrial democracies bound themselves in a pluralistic security community. The same concerns ultimately played a significant role in bringing the cold war to a peaceful end. Influential figures in both camps came to recognize the dangerous and counterproductive consequences of arms races and the sustained competition for unilateral advantage. With Gorbachev acting as a catalyst, the superpowers transformed their relationship and, by extension, the character of the international system.

To the extent that actors can, wittingly or unwittingly, change the 'rules of the game,' and even the nature of the political and economic systems in which they operate, general theories of process in international relations will have restricted validity. Unlike theories of evolution, they will not apply to all of history, but only to discrete portions. It seems self-evident but needs to be emphasized: scholars need to specify carefully the temporal and geographic domains to which their theories are applicable. We suspect those domains are often narrower and more constrained than is generally accepted.

A second big difference between international relations and evolutionary biology is the purpose of the endeavor. International relations scholars cannot predict the future, but neither can we ignore it. People need to make decisions in the face of uncertainty about the future, and consequently they need appropriate concepts and foci for information to maximize the quality of those decisions. As deductive-nomothetic theory is of very limited utility for this purpose—something policymakers have known for a long time—scholars need to develop some other, more useful method if we are to have any influence as a profession on important policy dilemmas.[6]

Policy-relevant social science considers the general *and* the particular and goes back-and-forth between them to make sense of social reality.[7] At the general level, we have numerous (if fundamentally untestable) propositions and less formal understandings of some of the conditions in which war and peace may be more likely to occur. With regard to war, historians and social scientists alike have distinguished between need- and opportunity-based resorts to force and have identified different sets of conditions associated with each. These include but are not limited to general power capabilities, the military balance between states and alliances, expected shifts in any of these balances, and domestic problems that threaten leaders, regimes, or states themselves. More broadly, decisions to use force also appear to be influenced by the general state of regional and international affairs, dominant moral and intellectual conceptions, arid salient historical analogies. We need to treat all these factors as defining possibilities in particular circumstances; but no combination of them; can predict what choices real actors will make.

Take the example of the post-'victory' conflict in Iraq, which one reviewer of this chapter objected was quite predictable. What this objection ignores is how senior administration officials involved in war planning systematically sidelined such predictions and the implications of that choice for how events would unfold. The specific decision on troop deployments nicely illustrates the problem for social research. Extensive debate and analysis within the Pentagon, CIA, and National Security Council produced widely varying estimates of the deployment needed in postwar Iraq depending on what peacekeeping 'model' of troops to population was

---

[6]George (1993) makes a similar point.
[7]Carlsnaes (1992, 1993) has made a similar argument.

used as a baseline. According to a confidential NSC briefing for Condaleeza Rice, "Force Security in Seven Recent Stability Operations," the Kosovo model 'predicted' the need for 480,000 troops in postwar Iraq, compared to 364,000 for the Bosnia model and only 13,900 for the Afghanistan model (Gordon 2004). No combination of factors could have predicted that Defense Secretary Donald Rumsfeld would later dismiss the higher estimates. Even if existing explanations for post-conflict conditions had been sufficient to demonstrate Rumsfeld's poor judgment, the actual decision to ignore those estimates produced significant |and unanticipated outcomes as events unfolded on the ground. The chaos! in many cities following the fall of Baghdad, for example, created a hospitable environment for the nascent insurgency to establish a foothold in unsecured areas, organize itself, and gain public support. Even in this 'predictable' case, a research strategy that identifies early indicators of which model (if any) of post-conflict peacekeeping is playing out or provides warnings of the need to revise troop estimates as events unfold is of greater utility than one that promises predictable outcomes.

Put more generally, concreteness requires culturally local knowledge, because states, ruling elites, and individual leaders respond differently to similar combinations of threats and opportunities. Incentives ultimately are in the eye of the beholder. Leaders may also respond differently to similar stimuli before and after experiences that transform their identities or their understanding of ongoing strategic interactions in which they participate. We need better tools to wed general knowledge about international relations and foreign policy to the more specialized knowledge that area and country experts have about actors in specific conflicts and contexts.

3.3 Forward Reasoning

The logic of our argument suggests that point prediction in international relations is impossible. Evolutionary biology is not a tool for explaining current 'trends.' It is at best a limited tool for identifying relevant trends, but not until fairly long after the fact, because such a multitude of forces and random interactions determine the course of evolution. As we have argued, social scientists cannot afford the luxury of only examining the past, they are deeply engaged in the attempt to explain the present and think analytically about the future. Our interest is in the identification and connection of chains of contingencies that could shape the future.

One useful approach is the development of scenarios, or narratives with plotlines that map a set of causes and trends in future time. This forward-reasoning strategy is based on a notion of contingent causal mechanisms, in opposition to the standard, neopositivist focus on efficient causes, but with no clear parallel in evolutionary biology. It should not be confused with efforts by some to develop social scientific concepts directly analogous to evolutionary mechanisms (such as variation or selection) in biology to explain, for example, transformations in the international system or institutions, or conditions for optimum performance in the international political economy.[8]

Scenarios are not predictions or forecasts, where probabilities are assigned to outcomes; rather, they start with the assumption that the future is unpredictable and tell alternative stories of how the future may unfold. Scenarios are generally constructed by distinguishing what we believe is relatively certain from what we think is uncertain. The most important 'certainties' are common to all scenarios that address the same problem or trend, while the most important perceived uncertainties differentiate one scenario from another.

This approach differs significantly from a forecasting tournament or competition, where advocates of different theoretical perspectives generate differential perspectives on a single outcome in the hope of subsequently identifying the 'best' or most accurate performer. Rather, by constructing scenarios, or plausible stories of paths to the future, we can identify different driving forces (a term that we prefer to independent variable, since it implies a force pushing in a certain direction rather than what is known on one side of an 'equals' sign) and then attempt to combine these forces in logical chains that generate a range of outcomes, rather than single futures.

Scenarios make contingent claims rather than point predictions. They reinsert a sensible notion of contingency into theoretical arguments that would otherwise tend toward determinism. Scholars in international relations tend to privilege arguments that reach back into the past and parse out one or two causal variables that are then posited to be the major driving forces of past and future outcomes. The field also favors variables that are structural or otherwise parametric, thus downplaying the role of both agency and accident Forward reasoning undercuts structural determinism by raising the possibility and plausibility of multiple futures.

Scenarios are impressionistic pictures that build on different combinations of causal variables that may also take on different values in different scenarios. Thus it is possible to construct scenarios without preexisting firm proof of theoretical claims that meet strict positivist standards. The foundation for scenarios is made up of provisional assumptions and causal claims. These become the subject of revision and updating more than testing. A set of scenarios often contains competing or at least contrasting assumptions. It is less important where people start than where they are through frequent revisions, and how they got there.

A good scenario is an internally consistent hypothesis about how the future might unfold; it is a chain of logic that connects 'drivers' to outcomes (Rosell 1999: 126). Consider as an example one plausible scenario at the level of a "global future" where power continues! to shift away from the state and toward international institutions, transnational actors, and local communities. The state loses its monopoly on the provision of security, and basic characteristics of the Westphalian system as vie have known it are fundamentally altered. In this setting, key decisions

---

[8]See, for example, Modelski/Poznanski (1996), and other contributions to the September 1996 special issue of *International Studies Quarterly*.

about security, economics, and culture will be made by non-state actors. Security may become a commodity that can be bought like other commodities in the global marketplace. A detailed scenario about this transformation would specify the range of changes that are expected to occur and how they are connected to one another. It would also identify what kinds of evidence might support the scenario as these or other processes unfold over the next decade, and what kind of evidence would count against the scenario or indicate a branching off point. Moreover; evidence that: counts against one scenario might count for another. For example, whereas a plotline that included September 11 might not have been anticipated, alternative scenarios that led to futures where the state reasserted its security function might have been constructed. Evaluations of new evidence as events unfolded would then determine which scenario appeared to be playing out, or whether, the same scenario had started to evolve in unanticipated directions. The same drivers could be at play in multiple scenarios, but how changes in technology, human agency, and transnational networks interact is less certain and these interactions can lead to outcomes along very different trajectories.

This method is simply a form of process tracing, or of increasing the number of observable implications of an argument, in future rather than past time. Eventually, as in the heuristics of evolutionary biology future history becomes data. But instead of thinking of data as something that can falsify any particular hypothesis, think of it as something capable of distinguishing or selecting the story that was from the stories that might have been. Such storylines should not be thought of as linear but as contingent in a way our scenario methodology tries to capture.

The scenario methodology has seven steps: identification of driving forces, specification of predetermined elements, identification of critical uncertainties; development of scenarios with clear 'plotlines,' extraction of early indicators for each scenario; consideration of the implications of each scenario, and development of "wild cards" that are not integral to any of the scenarios but could change the situation dramatically if they were to happen.

Driving forces are the causal elements that surround a problem, event, or decision. While some driving forces are likely to derive from standard causal arguments in major social science theories (e.g., the diffusion of power and the growth of commodities markets), others are not. In developing explanations for past events it is common to identify only a few, even two, driving forces. We call them "independent variables," which implies, of course, that they are somehow independent (of each other and of other causes). In generating scenarios the starting point is to put on the table multiple driving forces that can be the basis, in different combinations, for diverse chains of connections and outcomes. Parsimony comes after, not before, an analysis of complex causal possibilities.

Predetermined elements appear relatively certain. They are parameters that can safely be assumed for the scope and span of the scenario exercise. One goal of a scenario is to separate what appears certain, or very close to it, from what people simply think or believe is likely, without engaging in well-established psychological processes of treating routine events, 'causes') of 'effects' and 'structural' causes as immutable.[9]

There are no easy experiments and control situations in world politics, but we can still assert with confidence that some developments appear nearly certain. Examples include slowly changing phenomena, such as demographics, and constraints such as geography and physical resources. We nevertheless need to be very careful in categorizing elements as certain.

In the 1970s, the experts assumed that oil reserves were rapidly becoming depleted, only to be surprised by new discoveries. It| seems reasonably safe, however, to assume that new water will not be discovered in the Middle East, and that limited supplies constitute a real source of friction between Turkey and Syria, and between Israel and Palestine. We must be even more cautious about political 'certainties' and "social facts."[10] In the 1970s many theorists treated as given intense and ongoing conflict between Egypt and Israel, and between the United States and the Soviet Union. In both cases, scholars were profoundly surprised by the termination of these conflicts and the reshaping of the regional and international environments that resulted.

Critical uncertainties describe important determinants of events whose character, magnitude or consequences are unknown. This uncertainty can also be the result of unknown interaction effects among combinations of the predetermined elements. Scenarios highlight the critical uncertainties; the plotlines confront these uncertainties directly as connecting principles that pull the story together.

Standard social science theory 'testing' treats as mutable the "independent variables" suggested by connecting principles that we already know well. In scenario thinking, plotlines have to work with the critical uncertainties rather than the other way around. This is often a serious challenge, because it is impossible to know in advance of the empirical data what combinations of driving forces might come together in a setting of multiple conjunctural causality to yield particular outcomes. Of course, it is precisely that challenge that makes the scenario method a valuable tool. The goal is to learn from the future (as it unfolds), not predict it. No set of scenarios captures a comprehensive picture of all possible causal combinations—and it is not necessary to do so. What are necessary are clear causal relationships, even if complex. These can be evaluated, and modified, in response to emerging data.

A *scenario plotline* is a compelling story about how things happen. It describes how driving forces might plausibly behave as they interact with predetermined elements and different combinations of critical uncertainties. Plots have their own logic—sometimes more; than one logic—that drive the story forward and suggest the directions in which the uncertainties may resolve. The logic(s) may be drawn from standard international relations theories. For example, balance of power theory

---

[9]See 'Introduction' in Tetlock/Belkin (1996).

[10]Searle (1995), defines social facts as those facts produced by virtue of relevant actors agreeing that they exist. See also Ruggie (1998).

emphasizes the way in which a strong driving force (states (see footnote 5) desire for independence and autonomy) interacts with predetermined elements (power configurations) and critical uncertainties (who will ally with whom) in an international system to produce outcomes. But this is not the only logic applicable to international relations.

Competing theories or approaches identify different drivers and may lead to different behavioral expectations. Moreover, all these approaches acknowledge the importance— sometimes determining—of elements outside their theory, such as processes of diplomacy and personalities and preferences of individual leaders. The advantage of the scenario method is that stochastic events, equifinality, multifinality, and complex, conjunctural causation are no longer stubborn inconveniences that need to be minimized or simply ignored. They can be treated as natural and fundamental aspects of reality. This can be done by developing multiple scenarios, or scenarios with branching points, that capture the probabilistic nature of the arguments at play, without, however, having to attach essentially arbitrary probability estimates to the strength of particular 'variables' or different outcomes.

Plotlines draw on and ultimately depend upon the existence of regularities in social interaction, in world politics as elsewhere.[11] But they consciously place these regularities in a contextualized setting and thus make no claim to identify invariant ontological structures or laws.

Early indicators are observable and measurable attributes of die political situation that allow researchers to assess, as events unfold, the extent to which a scenario (or which part of a scenario) is coming to pass. Developing early indicators is an exercise in "process-tracing," extrapolated into the future. If a particular set of driving forces were to become most important and lead to a given scenario, what would be some of the early indications that events were indeed unfolding along that particular path and not along another? The strategy is a modified version of the simple idea of increasing the number of observables that differentiates one set of explanations from another in a verifiable way.[12] By doing so in future time, we reduce post hoc determinism and force ourselves to confront historical contingency in a creative manner.

Implications of scenarios are aimed explicitly at decision making and choice. One of the valuable consequences of thinking about historical contingency in a disciplined way is that it forces people who are going to make decisions to ask what they would do if they found themselves in—or heading toward—a world different from the one they expect. Theory-based prediction compels decision makers to make or justify a decision or strategy on the basis of a single-point forecast (at best, with a range of uncertainty around it) whose accuracy cannot be known until after the outcome is known. With scenarios, actors can evaluate decisions against the

---

[11]For an effort to save the 'scientific' explanation while doubting the usefulness of general laws for explaining social phenomena see Elster (1989). See also Brown (1984).

[12]See, for example, King et al. (1994): 19, 28–29.

most plausible scenarios in the current set and then evaluate the likelihood of these scenarios as their strategy unfolds.

Considering at once the behavioral implications of more than one scenario helps to clarify the stakes, risks, and uncertainties connected with any single course of action that an individual or a state might choose. In some situations policymakers may be able to adjust their strategies in response to information that indicates their expectations are not being fulfilled. In others they may be able to hedge effectively against several different scenarios. Tracking through the use of early indicators might also help leaders to recognize that their actions could be an important pivot or determinant of the kind of future that was likely to evolve. Obviously, a process like this that included early consideration of several plausible scenarios, and the different ways the critical uncertainties might combine, could have been very helpful to U.S. political and military authorities before they chose to launch a war to change the regime in Iraq in 2003.

Finally, designers of scenarios need to consider *wild cards*. These are conceivable, if low-probability, events or actions that might undermine or modify radically the chains of logic or narrative plotlines. They might include assassinations, dramatic economic changes, and famines and natural disasters. Some wild cards could constitute extreme values on a familiar independent variable; others might be outside the realm of standard social science arguments. In either case, doing this prospectively could change our views on what variables should be a part of theory, or what an 'extreme' value actually is—since it avoids the possibility of post hoc certainty. It would also be revealing if we were to miss entirely a wild card type cause, or if wild cards happened but were "dampened out" in their effects by other kinds of causes.

A central choice in developing scenarios is whether to begin with drivers—the "causal forces" or the plotline in the story—or: the outcomes or resolution of the stories. There are several reasons to start with drivers. From the perspective of traditional social science, it is cleaner in principle to reason from cause to effect when possible. Pragmatically scenario thinkers are more likely to generate results that contain surprises or challenging combinations of events when they begin from beliefs or ideas about fundamental causes, rather than from preconceived notions of the most likely outcome states. People who work on particular problems and have done so for a long time typically carry around in their heads a set of plausible outcomes, or "official futures," that they believe are likely and relevant to their concerns. One of the purposes of constructing scenarios is to encourage scholars and experts to think outside of these confines about plausible, different futures.

In summary, scenario thinking is disciplined by beginning with the identification of the several factors (causes) that scholars believe are most important to the future of a political relationship. They can then distinguish between what is most certain and what is most uncertain. Uncertainty in this context can mean that scholars are uncertain about the 'value' of the variable, or about the causal impact of the variable, or both. The three or four most important uncertain causes can then be identified, as well as a narrative explication of the key uncertainties at play and the nature of their possible interactions. These critical uncertainties become the basis of different plotlines. By assigning different 'values' to these variables, and combining them in different ways, scholars can reason to a set of plausible end-states. These end-states should be plausible within existing conceptual frameworks, but, when possible, challenging to "official futures" Scholars can then develop the narrative pathways that could generate the outcomes by moving from a highly abstract framework toward increasingly precise—and compelling—causal stories that specify assumptions, major drivers, limiting conditions, and implications. As part of these narratives, scholars must specify the trends that weave through their stories and can be monitored as time passes.[13]

3.4   A Forward-Looking Research Agenda

The novelty of this approach in political research means we cannot draw from existing scholarship to support the fruitfulness of our approach. Nonetheless, we gain some confidence from the wide application of similar approaches in the policy sciences (e.g., health policy research and public health economics) and biological and physical sciences that share some of the epistemological challenges we have identified (e.g., climatology). Such approaches are especially ubiquitous when research fields have direct public policy implications. Moreover, empirical scholarship on expert political judgment suggests that decision making that is more consistent with the underlying logic of scenario-based strategies will produce better 'predictions' than deductive research strategies (Tetlock 2005). Specifically, longitudinal studies of expert predictions demonstrate that those with extensive specific knowledge, who draw on research from many fields, and are able to improvise and revise in response to changing events outperform those wedded to one tradition or who impose ready-made solutions.

In lieu of surveying existing research, this section applies the abstract understanding of a forward-looking research strategy to major trends in international relations. We do not elaborate full scenarios here.[14] Instead, we identify what we believe to be three of the most important developments likely to affect international relations in the coming decades: the continued increase in intrastate conflict, further proliferation of weapons of mass destruction, and an increasing privatization of security. We should note at the outset that we originally developed these plotlines in 2000, and have not changed our initial formulations so as to stay true to our methodology. Our purpose is to show how a forward-based method can be used to track, study, and understand these trends in a disciplined way. We make no claim that fundamental controversies in social science can be thus resolved, although we are confident that constructive forward-based thinking can help to clarify some of

---

[13]For a similar discussion of 'causality' embedded in a narrative explanatory protocol, see Ruggie (1998): 89–94.

[14]For how to construct scenarios, see Weber (1997) and Stein et al. (1998).

the parameters surrounding those controversies and the nature of the disagreements at hand.

Distinguishing trends, drivers, and outcomes can be conceptually difficult. The trends we identify may be outcomes caused by previous drivers, and also by drivers of other outcomes, most notably fundamental changes in the international system. Indeed, we chose the three I trends because we thought they were likely to contribute to important changes. The methodology of scenario construction has allowed us to monitor and revise our expectations. If indicators that we have specified with any one of these trends do not become apparent, we then reexamine underlying theoretical assumptions and reformulate the scenario. In this sense, the method is rather like an antiaircraft system, responding to feedback and readjusting its trajectory as history flies by.

Using scenarios as a research method, the goals of research expand to include not only the development of better explanation's, but also identification of points of intervention, ongoing revisions of scenarios as events unfold, and the consideration and reevaluation of salient causal pathways. Scenario methodology also highlights how learning and feedback may change possible futures in dynamic ways difficult to anticipate. This research strategy could easily be applied to particular regions— South Asia, the Middle East—or to particular relationships. We chose instead to focus on trends that cut across regions to show the most general application of the research strategy.

3.4.1 Intensified Ethnic Conflict

For the most part, the most violent and pervasive conflicts in the post-cold war period are within states, not between them. They nonetheless often become international when they spread across borders or draw in third parties as participants, would-be mediators, or peacekeepers. While a great deal has been written on specific intrastate wars and the general trend away from interstate violence, deductive theory has made relatively little headway in explaining within-state conflict or in understanding how to prevent its eruption. In part, the problem stems from inattention. During the Cold War, theories of international politics developed concepts and categories centered on states and strategic relationships, which said little or nothing about ethnic and civil conflict Despite this inattention, however, these conflicts were frequently an important foreign policy concern, a central contributor to superpower conflict, and a prominent item on the agendas of consumers of the resources of international institutions. A complicating factor is that the latest round of ethnonationalist conflict is occurring in an historical, strategic, and institutional context markedly different not only from that of the last fifty years, but also from the context of previous historical periods when such conflict was more common.

Research outside of international relations has uncovered a wide range of causes of intergroup conflict and violence.[15] These usually focus on local conditions that may cascade toward or trigger conflict ancient hatreds, manipulation by belligerent leaders, or fear-driven local security dilemmas between ethnic groups in the same territory.[16] Despite recent attempts by international relations scholars to incorporate these causes into their theories, complex interactions among a changing international institutional environment, relationships among major powers, and evolving local conditions create a formidable challenge. For reasons we have made clear, deductive theories are unlikely to capture the complexity of the interactions among the relevant factors at this stage in our understanding. Nor are they likely to predict communal conflict and, consequently, deductive theories will contribute little to prevention and to the limitation of human suffering produced by such conflicts. One recent review of ethnic and intrastate conflict literature termed mono causal arguments as theoretical "culs-de sac" that have pushed the "study of social violence into the same paradigm-level debates that have characterized the American study of international relations" (King 2004: 432).

A more modest and useful strategy would be to draw on past cases—Rwanda, Bosnia, Somalia, Sudan—to map the multiple paths to ethno-nationalist conflict, identify the contingencies and wild cards that played out, and construct several scenarios of communal conflict, each highlighting a different critical uncertainty.

Generalizing on the basis of the past is not enough. Conditions change, and belligerents may learn lessons, confounding the expectation that strategies that succeeded in the past will work in future conflicts. The lessons learned from Bosnia did not provide an adequate map for anticipating or responding to the crisis in Kosovo. Unanticipated responses, "wild cards" such as the accidental bombing of the Chinese embassy, and the complex interactions of local and external events require the consideration of new branches and new paths. Through scenario construction, analysts recognize that 'causes' may interact in unexpected ways and are sensitized to cues when events begin to track down alternative paths.

The scenario-building strategy begins with driving forces and traces through causal pathways as these drivers interact in specific circumstances.

Causal drivers of ethnonationalist conflict might include the breakdown of empires, the proliferation, evolution, and fragmentation of identities, and/or underlying demographic or environmental stresses caused by population growth and resource scarcity. The breakdown of empire is an example of a driving force derived from social science theory (Emerson 1960; Lasswell 1935; Kupchan 1994; Henderson and Lebow 1974).

When empires decay or collapse they can provoke intense conflicts by former minority groups attempting to create successor states. The competition of two or

---

[15]For a classic treatment of ethnic conflict see Horowitz (1985).

[16]For applications of various causal theories to the post-Cold War wave of ethnic violence by international relations scholars, see the series of articles in the Fall 1996 issue of *International Security* (Snyder/Ballentine 1997; Lake/Rothchild 1997; Ganguly 1997; Kaufman 1997).

more groups for the same territory has led in this well-known dynamic to some of the most intractable struggles of the twentieth century. The most acute variants involve successor states that have arisen from partition or have been subsequently partitioned. The end of the British Empire half a century ago left in its wake ongoing conflicts that include Northern Ireland, India and Pakistan, Greeks arid Turks in Cyprus, and the Israeli-Palestinian conflict. The collapse of the Soviet Union has generated similar conflicts along its former periphery—Armenia-Azerbaijan, Moldava—conflicts that give every indication of becoming intractable. The disintegration of Yugoslavia might also be considered a by-product of the Soviet collapse, with a smaller but intense set of conflicts associated with the breakup of a central state. Scenarios might be constructed that take early cues from postcolonial conflicts and the presence or absence of various local causes, but then consider Additional general and local drivers, in different combinations, to sketch out different plausible trajectories of conflict.

International norms are a more mutable driver that fall under the "critical uncertainty" category and thus need to be tracked. Sometimes they evolve slowly enough so that they can be treated as givens. However, they also may change rapidly, as many have following the cold war. Norms of humanitarian intervention are undergoing a particularly rapid period of evolution. Although sovereignty has never been absolute, the evolution of norms of intervention appears particularly uncertain as spheres of influence have disintegrated, global civil society has increased pressure on the international community to intervene when gross violations of human rights occur, and fear of mass migrations and spillovers of conflicts have increased. Since these norms remain uncertain and not deeply embedded in international institutions and structures, it is impossible to predict on the basis of those norms which crisis will evoke an internatipnal humanitarian response, or whether that response will appear justified or convincing enough to be successful or sustainable. Alternative scenarios would weigh this humanitarian impulse differently and explore different catalysts.

The interaction of leaders and domestic politics with changing international norms is even more contingent. The "Somalia Syndrome," for example, 'taught' U.S. leaders not to commit ground troops in an unstable local environment and thus significantly affected subsequent | decisions in Haiti, Rwanda, Bosnia, and Kosovo. In May 1994 the jClinton administration issued new restrictive guidelines on humanitarian intervention; when the most intense genocide of the twentieth century began in April 1994, the United States stood aside and discouraged states and international organizations from timely and active intervention. One senior state department official, highlighting the problem of feedback and agency in paths to conflict, noted "It was almost as if the Hutus had read it [the guidelines]" (Weiss 1995: 172).

Recent work on the micropolitics of social violence reinforces the importance of reflexivity. Commenting on the central finding of Beissingers (2002) study of social mobilization and collective violence in; the collapse of the Soviet Union, a recent review notes that violence "cannot be understood, much less modeled, without taking account of the !reflexive power of mobilization itself" (King 2004: 441). Calling! Beissingers book "an elegantly theorized account of the power of contingency," King (2004: 444) notes that "[leaders'] calculus of costs and benefits, such as it was, was demonstrably influenced by their assessment of what had succeeded and filled in other circumstances. ... It was the very context in which individual events took place that accounts for how over time the impossible came to be seen as inevitable." In other words, the clustering of protests and social mobilization themselves shaped a future of j action against the state, the eventual collapse of the Soviet Union, and the social mobilization and solidarity that subsequently; in some local contexts, led to civil war (King 2004: 441).

The metaphor of disease, illness, and decline, initially suggested by Thucydides, and more recently by Bobrow for analyzing insecurity, fits nicely with our approach to forward reasoning. As Bobrow (1996: 446) puts it, "Implicit or explicit strategy recommendations should then carry warning labels. They also' should be subject to continuing monitoring for adverse consequences." They may have adverse side effects, and their use can sometimes produce immunities that make them ineffective in the future. For example, humanitarian efforts, peacekeeping, and safe havens in Bosnia may have prolonged conflict, and, by creating ethnic enclaves, even assisted Serbs in ethnic cleansing.[17] That is not to say that more forceful intervention might have had different results. Forward tracking and careful monitoring can help to expose where and why policies veer from anticipated trajectories and can highlight critical points of intervention as new 'branches' emerge.

Work on resource scarcity and acute conflict could also easily become the basis for a senario-based approach to intrastate conflict (Homer-Dixon 1999). Homer-Dixon maps the relationship between apparently unalterable trends such as demographic pressures or depletion of natural resources to their impact on local social and political conditions to produce potential conflict He argues that environmental scarcity constitutes an understudied set of causal variables that may be increasingly important as an underlying cause of acute violent conflict, although, he cautions.

The relationship between environmental factors and violence is complex. Environmental scarcity interacts with factors such as the character of the economic system, levels of education, ethnic cleavages, class divisions, technological and infrastructural capacity and the legitimacy of the political regime. These factors, varying according to context determine if environmental stress will produce the intermediate social effects [poverty inter-group tensions, population movements, and institutional stress and breakdowns]. Contextual factors also influence the ultimate potential for conflict or instability in a society. (Homer-Dixon 1996: 45).

Homer-Dixon's candid assessment of the limits of the causal claims of his research identifies many of the problems of research informed by the ideal of the covering law: uncertain relationships between underlying and immediate causes, open systems, complexity, negative degrees of freedom, and feedback and learning. The benefit of continually modifying and sharpening hypotheses in an effort to demonstrate that their validity is unclear when "the causes of specific instances of

---

[17]For a discussion of feedbacks and unintended consequences of interventions in the former Yugoslavia see Pasic/Weiss 1997.

violence are always interacting sets of factors, and the particular combination of factors can vary greatly from case to case" and are "often unique to| the society in question" (Homer-Dixon 1999: 7,178).

A more pragmatic and effective approach would be to begin with the same causal variables Homer-Dixon identifies, but work with the assumption that these multiple possible causes of environmental scarcity, including constrained agricultural productivity, migrations, and social segmentation, can interact in unanticipated ways with unexpected contingencies to complicate the paths to conflict and create new branches. What is critical is a well-specified set of indicators that can track 'evolution.'

These putative causal linkages are 'emplotted' storylines that can be analyzed in particular cases[18] but require sensitivity to feedback, interventions, surprises, or "wild cards" and the recognition that other drivers are equally plausible. Alternative scenarios should even consider different causal logics that stem from the same drivers. For example, a growing body of empirical work suggests that resource abundance in poor countries, not scarcity, can lead to conflict because it can "motivate rapacious behavior and allow the finance for civil war... It has been common knowledge that many of today's most durable conflicts, such as Angola, Liberia, the Democratic republic of Congo, Sierra Leone, etc. are fueled by the struggle for control of oil, diamond mines, timber and other resources" (de Soysa 2002: 7). Refinement and validation of hypotheses is unlikely, for reasons that we have made clear, to produce a definitive causal story that can be stated as a deductive explanation of a law. Similarly, no generic or off-the-shelf strategies of intervention or assistance are likely to prevent trajectories that appear to be moving down the path toward conflict. But a forward-reasoning approach could assist leaders. Context-specific scenarios could provide early warning of dangerous trends and sensitize analysts to local contingencies. Leaders could become aware of the plausibility of more than one future, design strategies of intervention and test these strategies for robustness and adaptability against different scenarios.

3.4.2 Nuclear Proliferation

The unexpected nuclear tests in India and Pakistan in the spring of 1998 quickly altered the security environment in South Asia and beyond. While proliferation of nuclear weapons—or weapons of mass destruction, more broadly—has neither been as uncontrolled or as limited as pessimists or optimists predicted, the explosions in South Asia highlight the importance of contemplating multiple causal pathways and multiple implications in the face of uncertainties and "wild cards."

---

[18]Polkinghorne (1988: 19–20) uses the literary term 'emplotment' to describe causation embedded in narrative: "It is not the imposition of a ready-made plot structure on an independent set of events; instead, it is a dialectic process that takes place between the events themselves and a theme; which discloses their significance and allows them to be grasped together as parts of one story." Cited in (Ruggie 1998: 94).

The nuclear tests pose serious conceptual and policy challenges (Stein 2001). Many causes of proliferation have been suggested. While strategic environments matter—no state without serious enemies has proliferated—many states with enemies have not (Argentina, Brazil) while others have relinquished nuclear weapons (South Africa, some former Soviet Republics). There is a diverse set of explanations of why states choose not to develop weapons: the effective use of carrots and sticks by major powers; the power of 'taboos' on weapons of mass destruction; and decision-makers' specific calculations of whether the risks from increased security dilemmas or being a possible target of preventive war outweigh the possible gains of nuclear weapons. In addition to explanations that focus on external calculations, domestic political factors increasingly appear important as well. For example, Solingen (1994, 1995) argues that liberalizing domestic coalitions, as opposed to more nationalistic or fundamentalist coalitions, are more likely to favor nuclear disarmament in order to strengthen the international economic ties on which they rely.

Separate from the puzzle of proliferation itself are the competing analyses of the consequences of proliferation, both regionally and globally. A number of broad-brush scenarios of possible futures already exist in the literature. In the 1960s, Herz (1968) wrote of "neo-territorial future in which sovereign states recognize not only their interests in mutual respect for each other's independence but also the need for extensive cooperation. Herz argued this kind of cooperation would become possible when the danger of nuclear destruction made all people and societies on the globe recognize their interdependence and their common fate. Interestingly, this scenario was a revision of his earlier argument that the nation-state would decline with the advent of nuclear weapons technology (Herz 1957). The evolution of Herz's thinking is very much consistent with the forward reasoning approach we propose: he recognized that the causal driver of nuclear technology produced unanticipated consequences and interacted with nationalism and state legitimacy in unanticipated ways. The outcome was retrenchment rather than demise of territoriality.

Following a similar logic that may resonate even more as proliferation progresses, Deudney (1995a, b, 2000) has presented a functional theory of how the international system might evolve into a global "Philadelphia System," similar to the governance arrangement that he argues prevailed in pre-Civil War United States, 1787-1865. He calls this "negarchical republicanism, residing between anarchy and hierarchy, where certain functions such as the control of nuclear arsenals might be embedded in cooperative institutions or multiple actor command systems while territorial units might maintain authority over other functions.[19] Although Herz's and Deudney's scenarios appear far distant in the future, creative

---

[19]Wendt (2003) goes further, proposing a teleological explanation for an inevitable world state. His logic follows in part on the same technological argument concerning the increasing capacity for devastation of military technology, but he also introduces an argument that the logic of anarchy will channel struggles for recognition. While such a well-developed plotline fits nicely with our scenario methodology, the functional and teleological logics on which Deudney's and Wendt's arguments are based run counter to our approach.

and disciplined thinking of this kind pushes forward the conceptualization of causal drivers that might lead to these outcomes and helps us to assess if and when we are on such a path. It would be worthwhile for Deudney to build in additional drivers and uncertainties to assess factors that open up or close off ways to get to preferred futures. Deudney might be able to argue that such a "negarchical republican response is functional (and rational) for human survival, but that does not mean it will occur. What are the links between our world and Deudney's future? Are international organizations such as the International Atomic Energy Agency developing the capacity or legitimacy to play the role that would be necessary for such a future to unfold? What would have to occur for established nuclear powers to give up full independent control?

There are also nuclear optimists such as Mearsheimer (1990) who argue from neorealist premises that proliferation will produce greater stability in a multipolar world. More nuanced studies also propose starkly different scenarios, as the debate on the pages of *International Security* between 'optimists' and 'pessimists' on the effects of proliferation attests.[20][19] It is worth noting that with the exception of the apparent certainty displayed by Waltz and Mearsheimer, both optimistic and pessimistic scholars recognize that "nuclear strategic logic is occasionally indeterminate or at least multifaceted, and … that many factors determine nuclear behavior" (Feaver et al. 1997: 186). Pessimists and optimists seem to agree that deductive theories based on the cold war experience are unlikely to apply as proliferation—or nonproliferation— proceeds. Sagan, a 'pessimist' about the effects of proliferation, harshly criticizes early post-cold war scholarship because it was dominated by "purely deductive arguments based on the logic of rational deterrence theory [that] eschewed the kind of historical research that is necessary to test theoretical arguments about the strategic effects of nuclear weapons" (Feaver et al. 1997, 193). Similarly, Feaver notes the need to supplement theoretical reasoning about U.S. or Soviet strategic behavior during the cold war, with "attention to causal relationships that drive the real-world behavior underlying observed outcomes." Careful attention to contingency in context allows the drivers of Soviet and American behavior during the cold war—domestic politics, cognitive traps, tradeoffs inherent in command-and-control—to be embedded in scenarios of future proliferation, but "in some cases, revised as new data becomes available" (Feaver et al. 1997: 186). Karl, a critic of Sagan and Feaver, stresses the need "to go beyond rote arguments over whether proliferation is good or bad and undertake empirical investigations into the actual behavior of new nuclear powers" (Karl 1996/97: 119).

Scenarios of the consequences of proliferation can make use of and build in a very helpful tool—competing game theoretic models—to identify cryptic and possibly critical dynamics of an important real world problem. Multiple nuclear powers with potentially opaque nuclear strategies and uncertain command-and-control systems are unlikely to operate as the simplest classical models predict. Scenarios

---

[20]Feaver et al. (1997). On the general debate between optimists and pessimists, see Sagan/Waltz (1995).

might also highlight the different factors on the path to preventive war or *military* strikes in potentially unstable regions such as the Middle East Would an Israeli air strike against nuclear facilities in Iran today produce the same muted response as did the strike on Iraq in 1981? This is not a question that could be answered by analogy or inference from any general theoretical understanding of international relations. Playing out different scenarios through different game-theoretic models might highlight dangers of alternative strategies, unexpected consequences under different contingencies, and opportunities to reduce tensions.

Plotting trends is not simply a matter of identifying multiple drivers. At times, it may involve detecting a shift in the conceptual terrain itself. When the conceptual terrain does shift, new understandings of international relations make nomothetic deductive theories all the more problematic. A growing body of international relations scholarship has pointed to the shifting ground of sovereignty (e.g., Biersteker/Weber 1996; Strange 1996; Ruggie 1993; Kratochwil 1995; Caporaso 2000). Identification of this conceptual shift has, however, had very little impact on mainstream 'scientific' theories, largely because analysis of "international conflict" rests on a Weberian conception of the state as the monopolizer of force.

The capacity to provide security as a public good to citizens has been both constitutive and defining for the modern state.[21] It has been constitutive insofar as war-making by the state directly and indirectly expanded its capacity to provide other public goods at home, and it has been defining in the sense that citizens gave their loyalty to the state as| their most important shield (Tilly 1975). The most far-reaching implications for the future of international relations stem from the possibility that (this understanding of the state no longer applies. This reformulation of the role of the state suggests that private security, supplied by the market, grows in relative importance, to public security supplied by the state.

Drivers of such a trend can already be identified. For example, in many parts of the world, fear of nuclear and even conventional war has declined precipitously. Citizens, no longer seized with the fear of nuclear war, have begun to think beyond physical security and to shift their agendas from the public to the private. Herz's and Deudney's propositions also drive in the direction of disaggregation of the security function of the state. Additional drivers include the effects of global markets, which put pressures on states to disengage as providers of other public goods; the [state becomes instead a regulator of the rules of the game or a supplier of competitive advantage.

New identities proliferate in such an environment As security from attack abroad becomes less of a preoccupation than it has been at any time in recent historical memory, the situational triggers that traditionally activate and affirm identification with the state are likely to decline in frequency. The disaggregation of security, when combined with the rise of an elaborate set of supranational institutions, may further disengage people from their connection to the state. If the state is not the

---

[21]For an analysis of the privatization of security and its consequences, see Stein (2000).

only supplier of security, its *command* of the loyalties of its citizens that separate them from the "other" in different countries may diminish.

Evidence already exists that drivers along this path to privatization are active. For example, the capacity of the state to protect its citizens at home has also declined, although it has declined in different regional spaces for quite different reasons. In the United States, the rise of 'gated' communities with private security systems contained behind walls is quite remarkable. Many large institutions—banks, schools, hospitals, and universities—now use private security forces to secure their local populations. It is not the state that secures its more privileged citizens from violent attack, but privately organized and financed security systems that are available in the market. Even public security providers are being contracted to the private sector to augment budgets. At the extreme, in Moscow, for example, private suppliers of security serve organized crime even as the capacity of the state to protect its citizens crumbles. Such trends have wide implications. Private markets for security over the long term will advantage the affluent and diminish identification with the state across social boundaries. While borders of states become less important, divisions within society may deepen if markets rather than states provide security. Political identities will be reshaped over time by the declining importance of state borders, and the growing importance of boundaries for private security markets.

The privatization of security is not restricted to the emergence of markets to supply the needs of the affluent within post-industrialized societies. In the wake of the end of the Cold War and the decline of empire, the major powers have progressively disengaged from regions they no longer consider of central strategic interest. Caught in a security vacuum, weak states have fragmented, not only in parts of Africa, but also in the former Soviet Union and Latin America as well. In Colombia, the state military, private paramilitary forces, and several guerrilla organizations compete to provide contracted protection to multinational corporations. Some of these fragmenting states no longer have the capacity to provide security for their populations; on the contrary, civilian populations are deliberately targeted by competing militias that supplant the forces of the state.

As state capacity to provide security declines, and international institutions retreat from the challenge, private suppliers of security increasingly fill the gap. At times they are contracted by international institutions, more often by states who seek to augment their capacity to coerce their own populations, and at times by nongovernmental organizations who seek access to insecure and desperate populations that are being systematically victimized by predatory militias. Private security markets are expanding in the shadow of fragmenting states and unwillingness by the major powers and international institutions to supply security as a collective good (Stein 2000).

The privatization of security, if it continues to widen and deepen, is likely to reshape the role of the state and shift political identities in global political space. The state, no longer the exclusive supplier of security, becomes one among several focal points of political identity. Borders, no longer the only or even the most important shield against attack, are likely to become increasingly less important as anything but a juridical divide between states, while boundaries—cultural and social divisions among spaces—drawn by private security markets are likely to become more important. These boundaries will not be as stable as state borders were in the twentieth century, because they are constructed out of market allocation not political authority. Nor will private purveyors of security be the focus of the kind of political loyalty that states were able to command.

We are not suggesting that this future will come to pass, or that it is the only plausible future. We must consider, for example, that the events of September 11, 2001 may represent a branching point. Borders appear again to be increasingly important, states are reasserting their security function at home and abroad, and international institutions in the security realm appear under strain. Yet, the underlying driving forces identified above have not disappeared. Even in the post-September 11 context, state militaries continue to "contract out" a number of functions, and supranational organizations appear increasingly necessary in conflict zones such as Afghanistan and postwar Iraq, the frontlines of the "war on terrorism." The "wild card" of September 11 provides an opportunity to construct alternative plotlines. Will its effects simply be "dampened out?" How will the interaction of a predetermined element not considered in our original formulation—the networking of private transnational terrorism and its link to radical religious movements—interact with technological change as described above? These interactions can be the basis of alternative plotlines.

Rather than prediction, laying out such a scenario and its alternatives encourages students of international affairs to consider a range of drivers, to identify the critical uncertainties, to develop different plotlines by varying these uncertainties, and to develop indicators of different paths to monitor trends as they unfold. Just as counterfactual analysis is a useful tool for evaluating the strength of competing explanations and recognizing the contingency of outcomes that actually occurred, forward reasoning opens our analyses to the possibilities of alternative futures but forces discipline in tracing likely paths created by important drivers in combination with significant uncertainties.

This analysis of plausible futures suggests that the nature of the units, identities, and key characteristics of the system can change. At least two of the trends we have identified—continued increases in intrastate conflict and privatization of security—suggest the need to reconceptualize the basic units of analysis as identities and the nature of human agency change. Some scholars have begun this reconceptualization of political actors. Ferguson and Mansbach,[22] for example, note that polities command loyalties of individuals and groups, but they hasten to add that the sovereign state is only one of the many forms and identities polities have taken over the ages. Multiple identities—whether ethnic, national, religions, professional, class, or ideologically based—and competing pressures for people's loyalties are nearly always present. This kind of reconceptualization of political actors and identities provides another starting point for analysis otherwise closed off by

---

[22]Ferguson/Mansbach (1996). See also Hall (1998).

deductive theories that posit relationships between given (usually state or national) actors. Different scenarios can be developed using particular conceptualizations of polities or actors as starting points, with analysis of critical uncertainties folded into different paths and plausible outcomes. These scenarios can be monitored along with more 'conventional' scenarios to assess where unfolding events fit best and where the 'storyline' needs to be adjusted. Using feedback from unfolding events, we can develop better and more compelling narratives of the future as we proceed through the present.

3.4.3   Conclusion

Newtonian physics conceived of a world of clock-like regularities that could be discovered through deductive theory and empirical research. Prediction was a reasonable goal because many of the phenomena studied by 18th- and 19th-century physicists were the result of a few easily measurable forces or of interactions among an extraordinary large number of units that gave rise to normal distributions. Neither of these conditions prevail in international relations—or in much of modem physics.

Evolutionary biology is shaped by a multitude of forces and by quasi to fully random events whose interaction cannot be modeled. Evolutionary biologists do not aim at prediction but instead have focused their efforts on developing theories that explain the process and history of evolution. They have met with considerable success.

We believe that international relations is closer in its basic nature and amenability to scientific study, to evolution than it is to mechanics or fluid dynamics. Like evolutionary biology, most kinds of prediction in international relations are impossible. Theories of structure and process—if we had robust theories—would fail to capture some of the most critical factors responsible for political outcomes because, as in evolution, they would lie outside any of the theories.

Scenario-based forward thinking is a promising method for tracking the policies of actors and the evolution of the international system. Scenarios allow researchers to combine general knowledge of politics with expert knowledge of individual actors and situations, to build in context, complexity, variation, and uncertainty in the form of multiple narratives with numerous branching points, and to revise their expectations as events unfold. Repeated iterations of this process can reasonably be expected to improve the quality of our general knowledge of international relations, our ability to track specific developments and the outcomes that result, and our capacity to address the problems that these evolutionary tracks create.

Why scenarios? First and foremost because theorists and policymakers both need constructive ways to think about the future and parse out the uncertainties in an inherently unpredictable setting. This is necessary for intelligent action, but also for progressive improvements in theory-based understanding of world politics. The future is not predictable. Acknowledging the limits on prediction forces a theorist (concerned with the biggest questions in social science to deal first with the boundary conditions around any argument with efficient causes. As the recognized boundary conditions become more restrictive, which they are likely to rapidly do, contingent and complex causality comes to the fore.

Second, we propose scenarios because econometric models and logic cannot accommodate sharp discontinuities. Qualitative uncertainties—particularly uncertainties about the fundamental rules of the game or institutional structures—require a different type of thought process and evidence collection. Certainly there are theorists of international relations who maintain that there have been few sharp discontinuities in world politics over the last 300 years, but that position seems increasingly untenable to most There are huge risks, theoretical and practical, in attempts to fit incoming evidence to existing theoretical paradigms when qualitative discontinuities may be present Scenarios are one way to balance that risk.

A third reason to use scenarios is to provide a common vocabulary that helps to clarify the nature of disagreements. We have found in our work that a group of theorists generating scenarios about the future of the Middle East peace process divided along two dimensions of disagreement: contingent disagreements and fundamental disagreements.[23] Fundamental disagreements are the result of basic, almost primordial beliefs about the world and the nature of politics. These are probably irreconcilable by evidence. Contingent disagreements are the result of differences in beliefs about the boundary conditions under which certain relationships hold. Contingent disagreements can be gently pushed toward resolution with careful quasi-experimental research designs, but they first need to be identified. One of the key findings of our scenario process was that disagreements that theorists took at the start to be fundamental, were often revealed later in the process as contingent This is an important, if small, step on the road to cumulation.

Finally, scenarios are useful because the theoretical study of international relations needs new ideas and arguments just as much as it needs to test existing ones. We are not opposed to the disciplined, precise evaluation of hypotheses and theories that are adequately developed so that they are ready for this kind of treatment. We are concerned about a search for false certainty and about the relatively trivial nature, and lack of policy relevance, of many 'big' generalizations. Scenario thinking, obviously, is not a panacea for this problem. It is a complementary toolkit that has promise for generating new ideas and arguments, broadening the range of causal relationships that we study, and tracking the evolution of world politics through periods of discontinuous change, in ways that promise to better over time both understanding and action.

---

[23]See Stein et al. (1998).

References

Beissinger, Mark R. 2002. *Nationalist Mobilization and the Collapse of the Soviet State.* New York: Cambridge University Press.

Bhaskar, Roy. 1979. *The Possibility of Naturalism. A Philosophical Critique of the Contemporary Human Science. Brighton: Harvester Press.*

Biersteker, Thomas J., and Cynthia Weber. 1996. *State Sovereignty as Social Construct.* Cambridge: Cambridge University Press.

Bobrow, Davis B. 1996. "Complex Insecurity: Implications of Sobering Metaphor," *International Studies Quarterly* 40 (4): 435—450.

Bueno deMesquita, Bruce. 1981. *The War Trap.* New Haven: Yale University Press.

Caporaso, James A., ed. 2000. Special Issue: "Changes in the Westphalian Order." *Environmental Politics 2 (4): 1—34.*

Carlsnaes, Walter 1992. « The Agency-Structure Debate in Foreign Policy Analysis, *International Studies Quarterly* 36, pp. 245-70.

Carlsnaes, Walter 1993. «On Analysing the Dynamics of Foreign Policy Change: A Critique and Reconceputalizaiton *Cooperation and Conflict* 28, pp. 5-30.

Deudney, Daniel H. 1995b. "The Philadelphian System: Sovereignty, Arms Control, and Balance of Power in the American States-Union, Circa 1987-1861," *International Organization* 49 (2): 191-228.

Deudney, Daniel H. 1995a. "Nuclear Weapons and the End of the Real-State," *Daedalus* 124 (2): 209-31.

Deudney, Daniel H. 2000. "Regrounding Realism," *Security Studies* 10(1): 1-45.

Elster, Jon. 1989. *Nuts and Bolts for the Social Sciences.* Cambridge: Cambridge University Press.

Emerson, Rupert. 1960. *From Empire to Nation: The Rise to Self-Assertion of Asian and African Peoples. Boston: Beacon Press.*

Feaver, Peter D., Scott D. Sagan, and David J. Karl. 1997. "Proliferation Pessimism and Emerging Nuclear Powers," *International Security 22 (2):* 185-207.

Ferguson, Yale H. and Richard W. Mansbach. 1996. *Polities: Authority, Identities, and Change.* Columbia, SC: South Carolina University Press.

Ganguly, Sumit. 1997. "Explaining the Kashmir Insurgency: Political Mobilization and Institutional Decay," *International Security* 21 (2): 76-107.

George, Alexander L. 1993. *Bridging the Gap: Theory and Practice in Foreign Policy."* Washington DC: US Institute of Peace Press.

Gordon, Michael R. 2004. "Catastrophic Success: The Strategy to Secure Iraq did not Foresee a 2nd War," *New York Times* (October 19), online edition.

Gould, Stephen Jay. 1989. *Wonderful Life.* W.W. Norton.

Gould, Stephen Jay. *1996. Full House: The Spread of Excellence from Plato to Darwin. New York:* Harmony Books.

Hall, Rodney Bruce. 1998. *National Collective Identity: Social Constructs and International Systems. New York: Columbia University Press.*

Harr A Rom, and Peter Secord. 1973. *The Explanation of Social Behaviour.* Oxford: Basil Blackwell.

Henderson, Gregory, and Richard Ned Lebow. 1974. "Conclusions," in Gregory Henderson, Richard Ned Lebow, and John G. Stoessinger, eds., *Divided Nations in a Divided World.* New York: David McKay Company, pp. 433-54.

Herz, John H. 1968. "The Territorial State Revisited—Reflections on the Future of the Nation-State," *Polity* 1 (1): 11-34.

Herz, John H. 1957. "The Rise and Demise of the Territorial State," *World Politics* 9 (4): 473-93.

Homer-Dixon, Thomas. 1996. *"The Project on Environment, Population and Security: Key Findings of Research," The Woodrow Wilson Center Environmental Change and Security Project Report 2: 45-48.*

Homer-Dixon, Thomas. 1999. *Environment, Scarcity, and Violence.* Princeton: Princeton University Press.

Horowitz, Donald L. 1985. *Ethnic Groups in Conflict.* Berkeley: University of California Press.

Jervis, Robert. 1997. *System Effects: Complexity in Political and Social Life.* Princeton, NJ: Princeton University Press.

Karl, David J. 1996/97. "Proliferation, Pessimism and Emerging Nuclear Powers," *International Security* 21 (3): 87—119.

Kaufman, Stuart 1997. "Spiraling to Ethnic War: Elites, Masses and Moscow in Moldova's Civil War," *International Security* 21 (2): 108-38.

King, Gary; Keohane, Robert O.; Verba, Sidney. 1994. *Designing Social Inquiry*; Brady and Collier, *Rethinking Social Inquiry..*

King, Charles. 2004. "The Micropolitics of Social Violence," *World Politics* 56 (3): 431-55.

Kratochwil, Friedrich V. 1995. "Sovereignty and Dominium: Is there a Right of Humanitarian Intervention?" in Gene M. Lyons and M. Mastanduno, eds., *Beyond Westphalia? State Sovereignty and International Intervention. Baltimore, MD: Johns Hopkins University Press, pp. 21-42.*

Kupchan, Charles A. 1994. *The Vulnerability of Empire.* Ithaca; Cornell University Press.

Lake, David A. and Donald Rothchild. 1997. *"Containing Fear: The Origins and Management of Ethnic Conflict,"* *International Security* 21 (2): 41-75.

Lasswell, Harold D. 1935. *World Politics and Personal Insecurity.* New York McGraw-Hill.

Lebow, Richard Ned, and Janice Gross Stein. 1990. "Deterrence: The Elusive Dependent Variable," *World Politics* 42: 336-69.

Lebow, Richard Ned. 2000-2001. "Contingency, Catalysts and International System Change," *Political Science Quarterly* 115: 591-616.

Mearsheimer, John J. 1990. "Back to the Future: Instability in Europe after the Cold War," *International Security* 15: 5-56.

Modelski, George, and Kazimerz Poznanski. 1996. "Evolutionary Paradigms in the Social Sciences," *International Studies Quarterly* 40:315-319.

Morris, Simon Conway. 1998. *The Crucible of Creation.* Oxford: Oxford University Press.

Pasic, Amir, and Thomas G. Weiss. 1997. "The Politics of Rescue: Yugoslavia's Wars and the Humanitarian Impulse" *Ethics and International Affairs* 11: 105-131.

Patomaki, Heikki. 1996. "How to Tell Better Stories about World Politics," *European Journal of International Relations* 2:105-34.

Polkinghome, D. 1988. *Narrative Knowing and the Human Sciences.* Albany: State University of New York Press.

Radelet, Steven, and Jeffrey Sachs. 1999. "What Have We Learned, So Far, From the Asian Financial Crisis?" Unpublished manuscript Rosell, Steven A. 1999. *Renewing Governance: Governing by Learning in the Information Age.* New York Oxford University Press.

Rudner, Richard. 1966. *Philosophy of Social Science.* Englewood-Cliffs, NJ: Prentice- Hall.

Ruggie, John G. 1993. "Territoriality and Beyond: Problematizing Modernity in International Relations" *International Organization* 47:139-74.

Ruggie, John G. 1998. *Constructing the World Polity. London : Routledge.*

Sachs, Jeffrey. 1998. "Global Capitalism: Making it Work" *The Economist* 12:23-25.

Sagan, Scott D., and Kenneth N. Waltz. 1995. *The Spread of Nuclear Weapons: A Debate.* New York Norton.

Searle, John R. 1995. *The Construction of Social Reality.* New York: Free Press.

Snyder, Jack and Karen Ballentine. 1997. "Nationalism and the Marketplace of Ideas" *International Security* 21 (2): 5-40.

Solingen, Etel. 1994. "The Political Economy of Nuclear Restraint," *International Security* 19 (2): 126-69.

Solingen, Etel. 1995. "The New Multilateralism and Nonproliferation: Bringing In Domestic Politics," *Global Governance* 1 (2): 205-27.

Stein, Janice Gross. 2000. "The Privatization of Security in Global Political Space," *International Studies Review* 2(1): 21-24.

Stein, Janice Gross. 2001. "Proliferation, Non-Proliferation, and Counter-Proliferation: Egypt and Israel in the Middle East,[5]" in Steven Spiegel, Jennifer Kibbe, and Elizabeth Matthews, eds., *The Dynamics of Middle East Nuclear Proliferation*. New York: Mellen Press, 2001: 33-58.

Stein, Janice Gross et al. 1998. "Five Scenarios of the: Israeli-Palestinian Relationship in 2002: Works in Progress," *Security Studies* 7:j 195-212.

Strange, Susan. 1996. *The Retreat of the State: The Diffusion of Power in the World Economy. Cambridge: Cambridge University Press.*

Tetlock, Philip E. 2005. *Expert Political Judgment How good Is It? How Can We Know?* Princeton: Princeton University Press.

Tetlock, Philip E., and Aaron Belkin, eds. 1996. *Counterfactual Thought Experiments in World Politics: Logical Methodological, and Psychological Perspectives. Princeton: Princeton University Press.*

Tilly, Charles. 1975. *The Formation of National States in Western Europe.* Princeton: Princeton University Press.

Waltz, Kenneth N. 1981. "The Spread of Nuclear Weapons: More May be Better," *Adelphi Papers* 178. London: International Institute for Strategic Studies.

Weber, Steven. 1997. "Prediction and the Middle East Peace Process," *Security Studies* 6:167-79.

Weber, Steven, 1996 "Counterfactuals Past and Future," in Philip A. Tetlock and Aaron Belkin (eds), *Counerfactual Experimens in World Politics.* Princeton: Princeton University Press. Pp. 268-90.

Weiss, Thomas G. 1995. "Overcoming the Somalia Syndrome—'Operation Rekindle Hope?'" *Global Governance* 1 (2): 171-187.

Wendt, Alexander. 2003. 'Why a World State is Inevitable: Teleology and the Logic of Anarchy," *European Journal of International Relations* 9:491-542.

Chapter 4
If Mozart Had Died at Your Age: Psychologic Versus Statistical Inference

Richard Ned Lebow

4.1 Introduction

A presidential address in the form of a short story is a novelty, but not a frivolous one.[1] My tale, which takes place in an imaginary world in which neither World War I nor II nor the *Shoah* occurred because Mozart lived to the age of sixty-five, is intended to do something that a scholarly presentation could not: dramatize the tensions between what I call 'psychologic' and the laws of statistical inference. The former describes the various cognitive and motivational biases that can influence information processing and so often make estimates of probability and attributions of responsibility different from what so-called rational models would expect. Biases and heuristics of all kinds can and have been described and documented by standard psychological studies and case studies of political decision making. Understanding biases intellectually and "feeling" them emotionally are not the same thing, and the latter, I contend, is necessary if we are to develop empathy for those who succumb to these biases and free ourselves from their grip.

This is especially true of the "certainty of hindsight bias." Baruch Fischoff has demonstrated that "outcome knowledge" affects our understanding of the past by making it difficult for us to recall that we were once unsure about what was going to happen. Events deemed improbable by experts (e.g., peace between Egypt and Israel, the end of the Cold War), are often considered 'overdetermined' and all but inevitable after they have occurred (Fischoff 1975; Hawkins/Hastie 1990). By tracing the path that appears to have lead to a known outcome, we diminish our sensitivity to alternative paths and outcomes. We may fail to recognize the

---

[1]This text was originally published as: "If Mozart Had Died at Your Age: Psychologic versus Statistical Inference," *Political Psychology* 27 (April 2006), pp. 157–72. Wiley grants the permission to authors to republish their texts "in a new publication of which [the author is] the author, editor or co-editor. This permission is here gratefully acknowlwedged.

© The Author(s) 2016
R.N. Lebow (ed.), *Richard Ned Lebow: Major Texts on Methods and Philosophy of Science*, Pioneers in Arts, Humanities, Science, Engineering, Practice 3,
DOI 10.1007/978-3-319-40027-3_4

uncertainty under which actors operated and the possibility that they could have made different choices that might have led to different outcomes.

Many psychologists regard the certainty of hindsight effect as deeply rooted and difficult to eliminate. But the experimental literature suggests that counterfactual intervention can assist people in retrieving and making explicit their massive but largely latent uncertainty about historical junctures, that is to recognize that they once thought, perhaps correctly, that events could easily have taken a different turn. The proposed correctives use one cognitive bias to reduce the effect of another. Ross, Lepper, Strack and Steinmetz exploited the tendency of people to inflate the perceived likelihood of vivid scenarios to make them more responsive to contin. People they presented with scenarios describing possible life histories of post-therapy patients evaluated these possibilities as more likely than did members of the control group who were not given the scenarios. This effect persisted even when all the participants in the experiment were told that the post-therapy scenario were entirely hypothetical (Ross et al. 1977). Tetlock/Lebow (2001) conducted a series of experiments to test the extent to which counterfactual 'unpacking' leads foreign policy experts to upgrade the contingency of international crises. In the first experiment, one group of experts was asked to assess the inevitability of the Cuban missile crisis. A second group was asked the same questions, but given three junctures at which the course of the crisis might have taken a different turn. A third group was given the same three junctures, and three arguments for why each of them was plausible. Judgments of contingency varied in proportion to the degree of counterfactual unpacking. There is every reason to expect that scholars exposed to counterfactuals and forced to grapple with their theoretical consequences will also become more open to the role of contingency in key decisions and events. So let the fun begin.

4.2   A Night at the Opera

The woodwinds sounded a minor chord as Oedipus fell to his knees. Blind, overcome by pain and grief, he remained motionless several bars after the sound of the woodwinds died away. Slowly and unsteadily Oedipus rose and stared with unseeing eyes at the audience. Two violins quietly began the famous fugue that ends Act III. When the cellos introduced the second voice, Oedipus groped his way towards the far right of the dimly lit stage and made his final exit.

The applause began before the last notes of the fugue had sounded. Erika rose from her seat, propelled by her enthusiasm for the performance, and through the act of clapping sought release from the tension of the last act. Hans, in the seat alongside, had been deeply moved too but could not bring himself to express his feelings so unguardedly. He envied Erika her ability to do so. Three curtain calls later, her tension spent, Erika turned to Hans who took her hand and guided her gently through the crowd toward the exit.

The night was brisk but not uncomfortable, and the couple, still holding hands, walked down the *Unter den Linden* in the direction of the *Brandenburger Tor.* They stopped for a traffic light and Erika broke the silence between them.

"Wasn't I right?"
"It was stunning," Hans agreed. "I'm glad we went."
"What did you think of Sussmann?"

The pedestrian signal turned green, and Hans led Erika into the intersection. He waited to answer her question until they were safely across the street. "He's the perfect Mozartian hero. He has a powerful but controlled voice. I thought he made the transition from fiery youth to mature statesman very convincingly."

Erika nodded. "As I see it, *Oedipus* is about the parallel conflicts between man's desire to assert free will against the fate that he fears determines his destiny and the striving for personal recognition and the need to feel and integrated part of a community. Oedipus unwittingly violates a communal taboo and punishes himself for this transgression. I like the way Mozart has the baritone sing *stonato* to convey Oedipus' internal conflict, and Sussmann does it as well as anybody I've heard."

"I sing off tune all the time, *schatz.*"
'Not deliberately!'
Hans gave her hand a squeeze.

The couple strolled along the perimeter of the *Tiergarten.* At Hans' sugges, they entered a cafe and took a table within reach of the heat of the large, open hearth fire burning seasoned oak from one of the many forests surrounding the city. A waiter appeared and wrote down their orders: *cappuccino* for Erika, a *pils* for Hans.

Erika reached into her pursue, pulled out a pack of cigarettes, put one in her mouth and lit it with a small, gold lighter which had also been stashed in her purse. Using both hands, she pushed her long blond hair back from her angular face. Hans recognized the gesture as a warning sign that Erika had something serious on her mind. He was nevertheless surprised when she leaned forward to ask what the world would have been like if Mozart had died a young man.

"What would the world be like?" Hans repeated her question.
"Yes. Would life be any different like today? Would Mozart's premature death have changed things in any way? Would we be sitting here having a drink?"
"I've never given it any thought. I certainly hope I'd still be sitting here with you."
"I hope so too." Erika reached across to give his arm a squeeze.
"If Mozart had died young...." Hans paused to consider the problem. "Well, we wouldn't have any of his mature works. There would be no *Oedipus,* no *Werther,* and no late piano and violin sonatas or symphonies, including my favorite, 57."
"True enough, but that's not what I had in mind. I was thinking about the broader artistic and political ramifications."
"I can see another one of your zany ideas is about to sally forth."
"You're just a stick in the mud, Hans. Suppose Mozart had been run over by a carriage when he was your age, 36. His last opera would have been *La Clemenza di*

Tito, a wonderful score to be sure, but an old-fashioned *opera seria,* and a far cry
from his mature, tragic works." Erika took a drag on her cigarette, and Hans lifted
his mug his beer to his mouth. He waited for Erika to continue, as he knew she
would.

"The post-classical movement would have been stillborn. Da Ponte, who wrote the
libretti for *Cosifan Tutte, Marriage of Figaro* and *Don Giovanni,* had to flee Vienna
after some count caught him in bed with his wife. Mozart did not find an equally
gifted collaborator until he teamed up with Neuman in 1805. That's when he really
began to explore the meaning of social justice and what kind of society would allow
man to reconcile competing needs. *Oedipus, Orestes,* and *Luisa Miller* are vehicles
for this analysis. Each opera examines some aspect of the problem in a more
complex way, intellectually and artistically."

"Mozart wasn't alone in addressing these themes."

"That's my point, Hans. In the early nineteenth century, music was the most *avant
garde* of the art forms. Mozart's mature compositions established a philo framework not only for music, but for literature and art, and even politics. Schubert and
Mendelsohn's music, Schiller and Shelley's poetry, carried on and developed the
post-classical tradition. Without Mozart, romanticism would have dominated the
artistic life of Europe and the political consequences would have been profound and
frightening."

"Come on, Erika. I know you despise romanticism, but that's an extraordinary
allegation."

"Bear me out, Hans. Romanticism represents the untrammeled expression of
individualism. Man ruled by his emotions, egoistic, self-indulgent, and uncon with
the consequences of his actions for others. Beethoven, Byron, Siegfried—all the
romantic heroes—real and fictional, are like this. Political leaders who were
influenced by romantic ideals would have made conflict a virtue, compromise
suspect, and passion in public life something to reward rather than constrain and
this at a time when rapid economic development and social change were creating
great political strains. What if romanticism had become the crucible for strong
nationalist movements whose leaders were not above using violence to achieve
their goals? It would have been very difficult, maybe impossible, to accommodate
the various struggles for reform or independence that developed in the last century."

"Your imagination never ceases to amaze me. But your argument is not convincing.
All those mass movements were attempts by disenfranchised classes to gain a share
of political and economic power. The Repeal Movement in Ireland was the pro-
totype. In the 1840's, Daniel O'Connell raised the prospect of political separation to
compel the British to restore an Irish parliament with substantial autonomy from
London. Other oppressed groups followed suit and met with varying degrees of
success. The most serious confrontation was in the Habsburg Empire where the
leadership's rigidity, suppression of dissent, and the late devel of a middle class
threatened to unleash chaos. Fortunately, cooler heads prevailed. Under the aegis of
Germany, France, and Britain, a federal solution was worked out in 1878.
Admittedly, it provoked a backlash by the Empire's German and Hungarian

communities, who until then had all but monopolized political and economic power."

"Don't minimize our country's problems, Hans. Remember the *Vaterland- spartei* that arose in the second half of the century and mobilized support from groups who had been marginalized by the industrial revolution. Beckstein, its leader until the great scandal, blamed the Jews for all of Germany's problems. He drew substantial support from the lower middle and artisan classes and disaffected intellectuals. As I recall, the *Vaterlandspartei* captured close to 20 % in the 1894 elections; they were helped, of course, by the economic crisis. During the Great Depression we had a similar if less successful movement, led by that Austrian, Hitler."

"He was a real flake." Hans raised his right arm in imitation of Hitler's signature salute.

"Ask yourself Hans why Beckstein and Hitler were flashes in the pans."

"That's pretty obvious. The 1849 constitution created the framework for a stable, decentralized, and democratic Germany. All the German states except for Austria sooner or later joined the confederation. Prussia, the most powerful and least democratic, was ultimately compelled to reform its electoral system. After that, it was only a matter of time before political power passed from the *Junkers* to the liberal industrialists. The more serious problem was labor unrest in northern Germany, Silesia, and the Rhineland. But once again, compromises were worked out, difficult as they were, and social democracy moderated its demands. German corporatism became the model for most of the rest of Europe. Because our political system was widely accepted as legitimate, neither Beckstein nor Hitler made much headway, even in times of economic crisis and widespread unemployment."

"You get honors in history, Hans. Now let me return to my counterfactual. Suppose in the absence of post-classicism, romanticism had come to dominate the artistic and political culture of the nineteenth century? Movements for reform would almost certainly have been movements for independence, and their leaders would not have sought independence as a means to an end but as an end in itself."

"I'm not sure I follow you?"

Erika brushed the hair back from around her cheeks and lit another cigarette. "In the latter half of the nineteenth century, many political movements demanded for independence to extract political and economic concessions from governments. The threat usually worked, and the idea of independence, not very practical in most cases, was shelved. In Quebec, they still play this game. A unified Italy made sense, and Vienna accepted the inevitable, but think of what would happened if the Habsburg Empire had been divided into a half-dozen so-called national states. None of these entities would have been viable economically. True 'national' states would have been impossible in practice because the various language communities are so geographically intermingled throughout the region. Hungarians and Ruma would have laid claim to Transylvania, Czechs, and Germans to Bohemia, Poles, and Germans to Silesia, Italians, and Germans to Tirol, Slovenes, Croats, and Italians to Istria—*und so weiter*. The Habsburg successor states would have fragmented into still small units and fought one another over disputed territories. I shudder to think of the consequences, especially in the Balkans."

"It's a ridiculous predicament to contemplate, I admit. But you haven't answered my question."

"Sorry about the digression, *Schatz*. Have you ever read Herder?"

"We discussed him in *Gymnasium,* but I don't remember much. Some mys nonsense about a nation being the organic expression of the soul of a people." "That's the gist of it. The traditional concept of a nation included all of the inhabitants of a political unit organized in several estates. For Herder, a nation was a group of people who belonged to a specific language or cultural community, regardless of where they resided. They had a right—a duty actually—to organize themselves politically in a 'nation state.' Some of Herder's successors carried his dangerous notion a step further by arguing that nations competed in a Darwinian world in which only the fittest would survive. To be one of history's winners, a nation had to become strong and carve out a niche for itself at the expense of its neighbors and competitors. Imagine a Europe of nation states with foreign policies based on the narrowest calculations of self-interest!"

"It would have been pretty grim."

"That's putting it mildly! Even in Western Europe, so-called nation states would have included substantial minorities who might have faced all kinds of discrimination. In Germany, the obvious target was the Jews. From the time of crusades they were the scapegoat for anything that went wrong. Beckstein and Hitler tried with only limited success to arouse hatred of Jews, but they might have succeeded in a different Germany."

"Now you're really getting carried away, Erika! France may have been the first country to tear down its ghettos, but Germany went much further than any of its neighbors in eradicating age-old prejudices. By the end of the nineteenth century, Jews, their religious practices aside, were indistinguishable from other citizens. Every schoolchild learns about the contributions they made to the scien, cultural, and economic life of our country. They even gave us our sense of humor. We Germans are an enlightened and tolerant people. This is why Beckstein and Hitler's hate mongering fell on largely deaf ears."

"Don't be so smug! It didn't have to be that way. In the absence of post, the twentieth century could have turned out very differently. Germany, France, and Britain, the three great democracies, were an axis of stability in a Europe reeling from the consequences of rapid industrialization. Without a common liberal framework to unite them, the great powers could have been at each other's throats as they had always been in the past. Look at Asia. Modernization in the absence of a common political culture, mutual economic dependence, and accepted mechanisms for resolving international disputes led to a series of destruc wars. Romantic nationalism would have undermined the basis for interna collaboration in Europe at the same time it would have encouraged more aggressive postures by the great powers. If Russia, Austria-Hungary, and maybe even Germany, had adopted expansionist foreign policies to cope with their domestic problems, a major

European war would have been hard to avoid. Suppose Germany and Austria-Hungary had lost such a war. Afterwards, some nut like Hitler, who attributed the defeat to a Jewish-socialist conspiracy, might well have found a receptive audience."
"That's far-fetched!"
'Why?'
"Since the age of enlightenment, Europe has witnessed the steady advance of reason and progress. Education, science, and economic development have ban ignorance, superstition, and poverty to the remotest corners of the continent. These developments have deep, structural causes. They are not dependent on particular individuals. It's very unlikely that the premature death of any artist could have profound consequences for the cultural development of an entire civilization. And even if it did, the triumph of one mode of artistic expression over another could never have led to the kind of political consequences you describe. After a certain point, the development of a peaceful and prosperous Europe was all but inevitable."
"You're blinded by the 'hindsight' bias."
"The what?"
"Did you ever read any cognitive psychology at university?"
"Are you kidding? Architectural students with time to read? But everybody knows about the cognitive revolution. Even *Leute von heute* had a story about Tversky and Kahneman and their institute in Vienna."
"*Leute von heute?* I didn't know you read such trash, Hans."
"It was in my dentist's office."
"Did the article say anything about the research of Baruch Fischoff? He's one of Tversky and Kahneman's Polish colleagues?"
"I don't think so."
"Fischoff discovered the hindsight bias. He found that once an event occurs, people upgrade their prior estimate of its probability. They see the outcome as almost inevitable and become correspondingly insensitive to the role of contin. The hindsight bias is one of the most ubiquitous and best documented of all cognitive biases."
"And you think I've fallen victim to it!"
"*I do,* Schatz.
"Well, I think you've gone too far in the other direction. If small changes in the world can have such large effects then almost anything is possible. One more hit of caffeine and you'll tell me how different European history would have been if Emperor Franz Josef had lived to be an old man and his nephew, Franz Ferdi, had never ascended to the throne. I can see the argument now. No Franz Ferdinand, no reforms, acute nationality problems in Austria-Hungary, and there's your European war."
"Who's getting carried away now?"

4.3   Imaginary Review of Manuscript 98-248

The story examines a counterfactual world in which Wolfgang Amadeus Mozart died at the age of 36. As a result, Europe became increasingly unstable and fought two destructive wars in the twentieth century. At the end of the first war, Austria-Hungary fragmented into a half-dozen unstable, independent states. Germany, the other big loser, also had to cede territory. A more fanatic version of Beckstein came to power, persecuted Germany's Jews, and started a second, unsuccessful war to regain Germany's lost territories.

The story is certainly imaginative and reasonably well written, and the two protagonists, an opera-loving husband and wife, are engaging characters. All the action is in the first paragraph that describes the final scene of *Oedipus*. After the opera, Hans and Erika retire to a cafe where Erika poses the Mozart counterfactual and describes its political implications to a sensibly dubious Hans.

The counterfactual, like most, is unconvincing. I can best demonstrate this by "unpacking" its first several steps. The antecedent, Mozart's premature death in 1791, is an acceptable minimal rewrite of history because 35 was close to the normal life expectancy in the era before modern medicine. Step two, in which the author contends that post-classicism never appeared, is highly problematic. Artistic styles capture or crystallize a society's mood. It is possible—I think likely—that some other composer, artist, or writer, or combination of them, would have developed post-classicism in the absence of Mozart. Artistic movements, like other human innovations, arise when the time is ripe. Physics offers a good example. At least a score of scientists struggled to understand the deeper meaning of the Michelson-Morley discovery that the speed of light in a vacuum was constant, regardless of its direction relative to the motion of the earth. If Poincare and Lorentz had not come up with their theory of relativity, somebody else like Planck or Einstein almost certainly would have.

The most Mozart's premature death could have accomplished would have been to delay the emergence and ultimate triumph of post-classicism. At least some cultural historians of the period argue that Haydn, among others, was already moving in this direction and might have developed post-classicism without the help of Mozart. If so, then cultural history would have been put "back on track," and the consequences of Mozart's death would have been dampened down rather quickly. Today's world would not be precisely the same—as Hans rightly observes; we would not enjoy *Oedipus* and other late Mozart works—but, *gross modo,* the world would be the same in its general political and cultural outlines.

Let us nevertheless assume that post-classicism never developed and that romanticism became the dominant cultural movement. The author insists (step three) that its triumph would have had profound consequences for European politics. Romanticism's celebration of imagination over reason and expression over argument would somehow have transformed moderate reform movements into extreme, even violent ones, that sought to break up the great multinational states that spanned Europe from the Rhine to the Urals. On the face of it, it is far-fetched to attribute such consequential political changes to variation in artistic style.

The political evolution of Central and Eastern Europe was largely determined by economics. Industrialization and trade produced a large and prosperous bour, an educated and better-off working class and the widely recognized need for economies of scale. There were tensions between the aristocracy and the new, rising classes, and these tensions were a contributing cause of revolution in France. By the end of the nineteenth century, the economic benefits of industrialization were apparent to nearly everyone, and the transformation it wrought in the distri of income and education compelled changes in the political structure. Even Erika, who makes the counterfactual case, recognized the economic and political chaos that would have resulted from fragmentation of the Austro-Hungarian Empire. So did key political actors from all classes and language groups; it was the principal incentive for them, with Germany's assistance, to work out a more democratic and federal structure for the Empire. The Magyar aristocracy, the biggest losers from the political changes, ultimately recognized that they too had much to gain from a peaceful and prosperous Balkans. Enlightened self-interest motivated the political restructuring of Austria-Hungary. It had nothing to do with the music to which people listened or the novels they read.

Counterfactual antecedents are linked to their consequents by a series of steps. Each of these steps is a development that is supposed to follow from the anteced, and all of the steps are necessary to produce the consequent. The Mozart counterfactual contains at least six steps linking its antecedent (Mozart's death at age 36) to the hypothesized consequent (an aggressive German regime in the twentieth century). Because Mozart dies young, (1) post-classicism fails to develop; (2) romanticism emerges as the dominant form of artistic expression; (3) reform movements in central Europe become nationalist; (4) Austria-Hungary and Germany go to war to cope with domestic and foreign threats; (5) they are defeated; and (6) in the aftermath, an anti-Semitic dictator hell-bent on a revision war comes to power in Germany. Each of these counterfactuals assume other counterfactuals (e.g., romanticism becomes dominant because no other artist develop post-classicism; Austria-Hungary behaves aggressively in the Balkans because Germany encourages it to do so instead of pushing Austria to reach an accommodation with its various linguistic communities; the victors in the first war are short-sighted, and unlike their predecessors at the Congress of Vienna, dis the losers and then stand aside and allow an obvious madman to come to power in Germany and bully its neighbors). The probability of the consequent is the product of the probabilities of every step. If we grant a probability of 0.5 for each step—and that is generous—the overall probability of the consequent is a mere .016.

There is admittedly something arbitrary about determining the number of steps in any counterfactual. Like a fractal—think of an ever-longer coastline each time a map of it is enlarged to show more indentations—a counterfactual can usually be subdivided into an almost infinite number of steps. As a general rule, the more steps a counterfactual requires the lower its probability. But the smaller the changes in history introduced by any counterfactual, the greater its likelihood. Conceivably, the overall probability of a counterfactual might not change signifi as we break it out into more and more steps.

There is nevertheless a difference, difficult as it may be to identify in practice, between the minimal requirements of a counterfactual (as I have tried to describe for the Mozart counterfactual) and the enabling requirements of each of these steps. Let me illustrate this with a counterfactual of my own. Marcel was injured in an accident while driving to work yesterday morning. I maintain this would not have happened if he had not listened to music instead of the news on the radio while eating his breakfast. Marcel took a different route to work because he heard the newscaster announce that the highway he normally takes was bumper-to- bumper in traffic. My counterfactual has two fundamental steps: Marcel does not listen to the radio and therefore does not change his route. There are many enabling steps: Marcel cannot slip on the ice while walking to the car, his car must start, and he must adhere to his regular route even when he discovers the highway ramp is backed up. The first two conditions have high probabilities. The third does not. Marcel might have responded to the tie up by crossing town via an alternate route. But he would likely have reached his office without incident because he would have arrived at the intersection where the accident took place at a different time and would not have been hit by the truck that earlier had spun out of control. Adding these enabling steps does not significantly affect the probability of Marcel arriving safely at his office.

Despite the improbability of the Mozart counterfactual, many readers will still find it convincing. People respond positively to narratives, and numerous psycho experiments indicate that a story becomes more credible the more detail it contains. This is because probability judgments are not attached to events but to descriptions of events (Tversky/Kahneman 1983; Tversky/Koehler 1994). In a recent experiment, Tetlock/Lebow (2001) showed that counterfactual (unpacking" leads foreign policy experts to increase their estimate of the prob of historical events. The more details they provided about possible, alternative outcomes to 1962 Cuban Sugar crisis (triggered by the European Federation's decision to give trade preferences to Cuban sugar while imposing stiff tariffs on American marketed sugar) the more likely the experts considered these outcomes. Lebow and Tetlock made no attempt to manipulate the number of steps between antecedents and consequents, but there is no reason why experi could not be designed to do this.

The laws of statistical inference suggest that the probability of *most* com counterfactuals will be low, and almost all counterfactuals that hypothesize major changes in the course of history have multiple links between their anteced and consequents. Does this mean that history is impervious to manipulation by counterfactual thought experimentation? No, only that the past cannot be changed to produce at will some *specific* world at any temporal distance. There may be *many* alternative worlds in which great powers fight a war in the early twentieth century, but we cannot know with confidence what counterfactuals could generate them, and even less, the specific characteristics of the alternative worlds these counterfactuals, or combination of them, would have. Because many alter worlds are possible, the probability of producing any one of them is low. For the same reason it is all but impossible to predict the future. Imagine a group of scholars meeting in 1815 to consider the character of the world a hundred years hence. Any world they describe would depend on numerous intervening steps. From the vantage point of 1815, the world in which we live today had a vanish low probability.

When thinking about contingency it is useful to distinguish between *specific* worlds and general *sets* of worlds. A specific world, like our own, has many features, one of which is the absence of a great power war. More importantly for our purposes, it is the result of a particular pathway of history. Our world is one instantiation of the set of *all* possible worlds in which there was no European war. A world in which Mozart died a young man would be a different world but still, I believe, a member of this set. Any number of other counterfactuals might produce other members of this set. There is also a class of worlds in which the great powers *did* fight a war, and Tetlock and Lebow came up with some rea counterfactuals for producing such a world. None of their counterfactuals involve the arts, and some of them I think require more than minimal rewrites of history. The probability of producing a counterfactual world in either set of worlds is much greater than that of producing a specific member of either set. This is because there are more paths that lead to worlds in the set than to any one specific world.

The course of human events is admittedly more malleable than a superficial examination of probability would suggest. It may nevertheless be governed by something akin to Heisenberg's principle of uncertainty. There is an inverse relationship between the magnitude of change we want to produce in the world and our ability to know if any counterfactual(s) will produce this intended change. This is also true for changes we introduce in the real world to produce desired future consequences. The more radical the change (counterfactual or real), the more steps between the antecedent and its consequents, and the greater the temporal remove of these consequents, the more unpredictable the outcome of the experiment.

For the sake of simplicity, I assigned a 0.5 probability to every step in the Mozart counterfactual, but the probability of these steps will almost certainly vary. Some steps may be more likely than others. In a world in which "national move" demanded independence, it would have been difficult to have reached a political accommodation in Austria-Hungary, and thus, the probability of interna conflict in the Balkans would also have been high. But I consider highly improbable the prior step on which the political stasis of Austria-Hungary depends: the determining influence of culture on the goals of reform movements in Eastern Europe. If the probability of one or more steps of a counterfactual is low, as I contend they are in the Mozart story, the likelihood of the consequent will be close to zero. If we assign a 0.1 probability to this step, and retain the 0.5 for all the other steps, the overall probability of the consequent is an insignificant .0031. If we assign higher probabilities to this and other steps, say 0.75, the likelihood of the consequent rises to 0.178—still less than one in five. To raise the probability over 0.5, the average probability of each step has to be at least 0.9 (this gives a total probability of 0.53), and this is very unlikely. This simple thought experiment indicates that for multiple step counterfactuals the probability of a consequent is more sensitive to the probability of individual steps than it is to the number of steps; even one step with a low probability will reduce significantly the probability of the consequent. While chains are only as strong as their weakest links, multiple counterfactuals are weaker than their weakest links.

The second assumption I made in calculating probability is that every fork leads to a meaningfully different alternative world. A six-step counterfactual like the Mozart story would generate 64 alternative worlds—assuming each step had two forks and each of these steps had two forks, and so on out to six steps. Many of the steps in this counterfactual could have multiple (more than two) forks. In the absence of Mozart, romanticism might entirely, largely, or only partially dominate the cultural life of Europe—or not at all if some other artist developed a version of post-classicism. The war Germany and Austria lost could have had multiple outcomes in terms of its human and territorial cost, and the particular mix of the two would surely have affected the probability of a dictator coming to power afterwards. A six-step counterfactual could generate more than 64 alternative worlds. These worlds would all be different in at least some respects, but many of them would probably be the same with regard to the attributes that concern us. This is because "second order" consequences would lead some, perhaps, many, of the forks back to a few real or alternative worlds.

Counterfactuals track a specific chain of historical developments arising from an antecedent. Second order consequences are developments outside this chain that might also follow on the antecedent and that could affect the probability of the hypothesized consequent. There is a prize winning history of the Peloponnesian War that argues that Athens could have won a victory early on if somebody else other than Pericles had been in charge and had pursued a more aggressive strategy against Sparta. But it is also likely that without Pericles, the Athenian assembly would not have reversed itself and offered an alliance to Corcyra. Without this alliance, war would not have arisen, and the counterfactual would have been moot. If we do away with some individual or development, we may also create a "niche" that other individuals or organizations fill. Earlier I suggested that without Mozart, some other composer, perhaps Haydn, might have developed post-classicism and put cultural history "back on track." Second order consequences can also affect the significance of the consequent, even if it does occur. Suppose we invent minimal rewrites that reverse the outcome of the Battle of Poitiers in 732, allowing Muslim invaders to penetrate France. If Muslim kingdoms in northern Spain continued their internecine fighting, weakening their overall power and drive to expand beyond the Pyrenees, then the benefits of their military success at Poitiers might have been short-lived. Second-order consequences would rather quickly have restored Christian political and religious dominance in France.

Only determinists will insist that second-order counterfactuals will ultimately make alternative worlds converge with the real one. Karl Marx insisted that the triumph of liberal democracy and rising standards of living for the working class were inevitable consequences of capitalism. If Disraeli had not introduced social legislation in Britain, and if Henry Ford not pioneered profit sharing in America, other people would have done these things. Such claims are extreme, but they find an echo in biology where ever since Darwin it has been recognized that evolution produces morphological similitude because there is a best set of physical charac and strategy for grappling with the challenges of life. Diverse species have converged independently on the architectures and behavior most suited to avoiding predators and exploiting food resources (Morris 1998).

I think it reasonable to assume that societies, like organisms, have a limited number of stable states. Accidents of history, real or counterfactual, can move the path of history away from these states, but there will be strong pressures to bring them back to the original or another stable state. Viewed in this light, the Mozart counterfactual is wanting in two respects. For reasons I have already made clear, it is the wrong counterfactual to produce the desired political consequent: a great power war in the early twentieth century. But even if some more appropriate counterfactual could produce that consequent, other developments—second-order counterfactuals—would sooner or later have brought about the peaceful, devel, democratic and closely integrated Europe that we enjoy today.

4.4 Author's Reply

The reviewer completely misses the point. Of course, the Mozart counterfac is far-farfetched—that's the whole idea! Erika is Germany's answer to Rube Goldberg. She or I could easily have provoked a European war by rewriting snippets of political history; Lebow and Tetlock did this, and convincingly for at least some of the historians they surveyed. Erika wanted to show how small, seemingly insignificant changes in reality can have large, unanticipated conse across different domains. By doing this, she hoped to make readers aware of just how contingent, interconnected, and unpredictable the real world is. None of the more prosaic, and admittedly, more convincing, paths to war would have accomplished this goal.

The reviewer claims to refute the Mozart counterfactual on scientific grounds. I shared his arguments with Erika, and she dismissed them as *Quatsch* [poppy]. After a long digression about the near impossibility of multistep counter- factuals, the reviewer concludes that what really matters is the probability of their individual steps, not their total number. The really telling question—and here we agree with the reviewer—is how to determine the probability of any step of any counterfactual. For some counterfactuals, this is pretty straightforward. Epidemi have robust equations that describe the factors responsible for the spread of infectious diseases. They routinely conduct counterfactual experiments by altering the values of one or more terms of the equation to see how it would retard or facilitate the spread of particular pathogens. For political counterfactuals this is impossible; there are no general laws that we can apply to specific cases or probabilities that we can calculate by observing the outcomes of a large number of similar events. Probability is a guessing game, and Erika's guesses are at least as good as the reviewer's.

When we strip away all the pseudo-science, the reviewer's prejudice is exposed. He or she is a crude determinist, no different from Karl Marx. The telling lines are the claims that culture could never influence economics or politics, but that economics determines politics. Does the reviewer offer any evidence for these assertions? None whatsoever. Many prominent historians believe that ideas deter the fundamental structure of any society's economics and politics and that economics is a branch of political science devoted to analyzing how politics shape economic decisions. Surely, there is room for different points of views—even in biology. At least one biologist, the American, Stephen Jay Gould, argued for the determining role of accident in evolution. He insisted that if you could rewind the tape of life and run the program over again you would end up with a radically different set of organisms each time (Gould 1989).

The reviewer makes an admittedly good point about second-order conse. Changes in reality ripple through society in unpredictable ways—this is a core assumption of the story. In theory, some other composer could have invented post-classicism. I acknowledge the need to add a paragraph or two to consider this problem in general and to explain why Haydn was not close to becoming a post-classicist.

The Mozart counterfactual was meant to provoke and make readers think about the contingent nature of our world. If Erika had wanted to be more 'scien,' she could have invited numerous counterfactuals that did not reach so far back into history and involved only minimal rewrites to provoke a European war. The same Lebow and Tetlock the reviewer cites *ad nauseum*—given the French example, I'm willing to bet the reviewer was that jughead Lebow—developed 10 counterfactuals that could have led to a great power war in Europe in the first decades of the twentieth century. In their first counterfactuals, the Congress of Vienna in 1815 awards Prussia land in Silesia and the Rhineland, where the industrial revolution got an early start, and made possible Prussia's rise to great power status. Germany becomes unified under Prussian leadership, assumes an authoritarian character, and, later in the nineteenth century, pursues an aggressive foreign policy. The other counterfactuals are scattered throughout the nineteenth century, and several of them take place during or on the eve of the crisis that supposedly led to a European war. The more proximate the counterfactual to the war, the fewer the steps between antecedent and consequent. Lebow and Tetlock surveyed historians, and they found some counterfactuals more plausible than others, and some of them likely to have produced the 'desired' war. Most signifi, there was no correlation between their judgments and the number of steps these counterfactuals entailed.

I must warn you, Erika is very unhappy about this Luddite review and has threatened to write a story in which soccer, not baseball, becomes the most popular European sport in the late nineteenth century. Just think of the political conse! Police—even those with helmets and leather jackets—were loath to mess with protesting students and workers. A stone in the hands of a fastball pitcher can be lethal. Police vulnerability encouraged concession and compromise, which in turn helped facilitate democratic transitions in France and Germany. If soccer had become the rage, Europeans would have been good at kicking but terrible at throwing. You can't kick cobble stones at police from behind barricades. European police and political authorities would never have been intimidated by students and workers, and we might still be living under repressive, authoritarian regimes! Reviewers like this one might not have had the ability—or the freedom—to throw their metaphorical stones.

References

Fischhoff, B. (1975). Hindsight is not equal to foresight: The effect of outcome knowledge on judgment under uncertainty. *Journal of Experimental Psychology*, 104, 288-299.

Gould, S. J. (1989). *Wonderful life*. New York: W. W. Norton.

Hawkins, S.A., & Hastie, R. (1990). "Hindsight: Biased Judgments of Past Events after the Outcomes are Known," *Psychological Bulletin* 107, no. 3, 311-27.

Morris, S. C. (1998). *The crucible of creation*. Oxford: Oxford University Press.

Ross, L., Lepper, M. R., Strack, F., & Steinmetz, J. (1977). Social explanation and social expectation: Effects of real and hypothetical explanations on subjective likelihood. *Journal of Personality and Social Psychology*, 35, 817-829.

Tetlock, P. E., & Lebow, R. N. (2001). Poking counterfactual holes in covering laws: Cognitive styles and political learning. American Political Science Review, 95, 829-843.

Tversky, A., & Kahneman, D. (1983). Extensional versus intuitive reasoning: The conjunction fallacy in probability judgment. *Psychological Review*, 90, 293-315.

Tversky, A., & Koehler, D. J. (1994). Support theory: A nonextensional representation of subjective probability. *Psychological Review*, 101, 547-567.

Chapter 5
Texts, Paradigms and Political Change

Richard Ned Lebow

5.1 Introduction

The contributions to this volume are indicative of the growing interest in Hans J. Morgenthau.[1] The principal catalyst of this interest is surely the end of the Cold War and the re-thinking of IR theory it has encouraged, Morgenthau is not the only theorist to whom scholars have turned; Thucydides, Nietzsche, Carl Schmitt, E.H. Carr, and Hedley Bull, and the 'English School' more generally are all undergoing revivals. If we widen our horizons to comparative politics and political theory, we observe a similar phenomenon with respect to Emile Durkheim, Leo Strauss, and Michael Oakeshott among others.

These revivals point to the need to conceptualize the relationship between political developments and texts. It seems obvious that major political changes or transformations—and more about what a transformation is in a moment—can provide an opening to criticize current policies and the discourses that sustain them, or are thought to do so. Older texts can be important resources in this effort. They can be used to decenter dominant discourses and provide a starting point, even a degree of legitimacy, to new ones, Morgenthau has proven a useful vehicle for exposing the pretensions of American foreign policy and limitations of neorealism. In *Tragic Vision of Politics* I argue that he is one of a number of useful resources for

---

[1]This text was first published as: "Texts, Paradigms and Political Change," in Michael Williams, eds., *Reconsidering Realism: The Legacy of Hans J. Morgenthau in International Relations* (Oxford: Oxford University Press, 2007), pp. 241–68. ISBN: 9780199288625. The permission to republish this chapter was granted on 3 July to the author who also retains the copyright for the original text.

© The Author(s) 2016
R.N. Lebow (ed.), *Richard Ned Lebow: Major Texts on Methods and Philosophy of Science*, Pioneers in Arts, Humanities, Science, Engineering, Practice 3, DOI 10.1007/978-3-319-40027-3_5

building theories that eschew prediction in favor of explanation, and offer themselves to policymakers as frameworks for working their way through problems.[2]

Contributors to this volume have used him for these several ends, as well as a subject of study in his own right, By examining the influence of Aristotle on Morgenthau, Anthony E. Lang, Jr., demonstrates the latter's concern for the ethical foundations of foreign policies and the essential role of prudence (*phronesis*) in successful foreign policies, William E. Scheuermann and Chris Brown explore the relationship between Morgenthau and Carl Schmitt. They make us aw are that realism is at the intersection of diverse traditions and owes debts to multiple predecessors, some of them people most realists would, with good reason, prefer to distance themselves from Scheuermann and Brown highlight an orientation common to many realist (as opposed to the liberal texts since the time of Thucydides: the extent to which they look back with nostalgia on an earlier period of history and view the story of their era as a *Verfallsgeschichte* or narrative of decay Oliver Jütersonke also acknowledges the connection between Morgenthau and Schmitt, but reminds us that Morgenthau's thinking was equally influenced by other important thinkers, notably Hans Kelsen and Hersch Lauterpacht, and legal-political debates which they animated, Nick Rengger situates Morgenthau within the Tragic tradition, and uses Michael Oakeshott's review of *Politics Among Nations,* and subsequent correspondence with Morgenthau, to question that appropriateness and utility of the tragic metaphor in IR.[3] Richard Little examines the complexities of Morgenthau's understanding of the balance of power and its connections to current theoretical debates. Michael Cox interrogates Morgenthau and his theory from the vantage point of the post-Cold War era, and uses his analysis in turn to help us better understand that conflict. He shows the tensions between Morgenthau and realism on the one hand, and US Cold War foreign policy on the other, and argues that some of these tensions arise from the nature of realism itself Drawing on his book, Campbell Craig reviews Morgenthau's evolving beliefs about a world state, and the tensions between his devastating critique of international idealism and the emergent, if not full-blown idealism, of his later writings, Michael Williams contends that Morgenthau is even more relevant today because of the framework he provides us to engage and critique the international politics of neo-conservatism.

In the first section of this chapter, I examine the linkages between texts on the one hand and political and intellectual goals on the other. I lay out four ways in which texts can be used for these ends, and offer illustrations drawn from recent Morgenthau scholarship and the chapters of this volume I then extend my argument to look at the broader question of the relationship between texts, paradigm shifts,

---

[2]Richard Ned Lebow, *The Tragic Vision of Politics* (Cambridge: Cambridge University Press, 2004), Chaps. 5 and 7.

[3]Lebow, *Tragic Vision of Politics*, offers a defense of tragedy, as Rengger recognizes. See Tom Erskine and Richard Ned Lebow (eds.), *Tragedy and International Relations* (2011), for a more thorough evaluation of the concept of tragedy as applied to international relations.

and foreign policy. These processes are to some degree self-reinforcing because contemporary interest in texts helps to define their status, and their status in turn helps to determine which texts scholars turn to as resources. As paradigms and discourses rise and fall, the scholars who work within them receive more (or less) recognition and resources, which in turn can accelerate the shift in either direction. I conclude with some observations about what we can learn from a better understanding of the historical process of revisiting texts.

5.2   International Relations After the Cold War

In a landmark US Supreme Court case, Associate Justice Potter Stewart wrote: "I can't define what pornography is, but I know it when I *see* it."[4] Political transformation is similar. It is difficult to define, in part because it is all in the eye of the beholder. It differs from pornography in that a consensus often emerges ex post facto that a transformation has occurred. This was true of the Cold War, whose origins, and even more its longevity, were not so obvious at the time. Bipolarity as a concept was introduced by William T. R. Fox in 1944, but not adopted by Morgenthau until 1950.[5] It only came to be regarded as a defining feature of that epoch sometime in the 1960s. The end of the Cold War came as a surprise to almost everyone, but was widely recognized as a watershed in IR even before the Soviet Union collapsed. Fools rush in where angels fear to tread, and IR theorists have not hesitated to propose definitions of system transformation. In 1979, Kenneth Waltz made it the centerpiece of his theory of international politics, and insisted that such change could only come about by war.[6] He further maintained that bipolarity was more stable than multipolarity, and certain to be with us for the foreseeable future. Events proved him wrong on both counts. The Cold War ended peacefully, and bipolarity—if it ever existed—came to an end when the Soviet Union imploded.

As Mick Cox observes, these events provided an opening for critics of neorealism, and of realism more generally, to go after them in a spate of articles and books. Most of these initial attacks focused on the failure of realists—or anyone else for that matter—to predict the end of the Cold War. Realists were held particularly responsible for this failure because their frameworks discouraged scholars from even acknowledging the possibility of a peaceful end to that conflict. It focused attention on the military balance, which really did not change markedly

---

[4]*Jacobellis v. Ohio*, 378 U.S. 184, 84 S. Ct. 1676, 12 L. Ed. 2d 793 [1964].

[5]William T, R. Fox, *The Super-Powers* (New York: Harcourt, Brace, 1944); Morgenthau, *Politics Among Nations*, 270–8. For Morgenthau's uses of the term bipolarity, see *In Defense of the National Interest* (New York: Knopf) 1951), 45; *Politics Among Nations*, 2nd edn. (New York: Knopf, 1954), Table of Contents and 325.

[6]Kenneth N. Waltz, *Theory of International Politics* (Boston, MA: Addison-Wesley, 1979), 165–70.

until the collapse of the Soviet Union. It also downplayed the significance of internal developments not directly related to material capabilities. These neglected other factors—which included the gradual disillusionment of Soviet intellectuals and *apparatchiki* with communist ideology, rising ethnic tensions, growing desires for material goods, and domestic and inter-Republic politics—that turned out to be where all the action was.

Realism was also morally compromised by the events of 1990–9. Prominent realist practitioners (e.g., Henry Kissinger) and academics (e.g., John Gaddis) valued stability over human rights, and had made dear their willingness to sacrifice Eastern Europe towards this end. As early as 1992, they began to regret the passing of the Cold War because of the uncertain and unpredictable nature of the world that was emerging in its place.[7] The most extreme expression of pessimism, and arguably, of amorality, was the much-criticized warnings given by John Mearsheimer to Japan and Germany, urging them to acquire nuclear weapons in what he insisted would be a far more threatening multipolar world. Fortunately, their leaders paid no attention to the unsolicited advice of marginal academics.[8]

The end of the Cold War provided critics of realism with both an opportunity and need to go on the offensive. The need side of the equation had several terms to it. First and foremost for some was the need to dethrone realism as the reigning paradigm to make room for other approaches, and with their ascent, the possibility of directing positions and funding towards them that would otherwise have gone to realists. Critics also had political agendas. Liberals were committed to the European project and a self-regulating community of industrial powers. Constructivists favored more far-reaching political transformations premised on the seeming success of what Karl Deutsch called pluralistic security communities.[9] As Mike Williams points out, neoconservatives were committed to a transformative agenda of a different kind, and here too, realism stood in the way. Realists of all stripes maintained that anarchy was the defining characteristic of the international system, and that it was impossible and downright dangerous to pretend that war was not the final arbiter of international disputes. Liberals and constructivists, by contrast, thought it possible to escape from, or at least, to mitigate, the worst features of anarchy through a dense network of institutions or a robust international society. The liberal position had been well developed even before the end of the Cold War, while constructivism was still an emergent paragim. Constructivists and their

---

[7]John Lewis Gaddis. Toward the Post-Cold World, *Foreign Affairs*, 70 (1991), 102–22.

[8]Kenneth N. Waltz, The Emerging Structure of International Politics, *International Security*, IS: 2 (1993), 5–43; John J. Mearsheimer, 'Back to the Future: Instability in Europe After the Cold War', *International Security*, 15:4 (3990), 5–56; Mearsheimer, The Case for Ukrainian Nuclear Deterrent, *Foreign Affairs*, 72: 3 (1993), 50–66.

[9]Karl W. Deutsch, et al. *Political Community and the North Atlantic Area: International Organization in the Light of Historical Experience* (Princeton, NJ: Princeton University Press, 1957).

fellow-travellers rallied round Alexander Wendt's timely article alleging that anarchy was what states made of it.[10] The battlelines were drawn, and the first skirmishes were taking place.

In the last decade, attacks on neorealism and realism have all but disappeared from the leading journals, indicating that critics believe the battle has been won. When the issue of realists versus critics does surface in these journals, it is because realists have gone on the attack. A recent example is John Mearsheimer's E.H. Carr Lecture at the University of Wales, Aberystwyth, in which he alleges that British academics discriminate against realists. It was printed as the lead article in the June 2005 issue of *International Relations,* and inevitably provoked a series of rejoinders by prominent British IR scholars.[11]

Critics of realism and realist critics of neorealism have displayed increasing interest in earlier realist texts, works that pre-date neorealism and the large body of contemporary realist literature that focuses on questions of material capabilities and their implications for foreign policy. There has always been interest in some of these authors, especially Thucydides, who is widely acknowledged to be the father of IR, not just of realism. Beginning in the 1980s, IR scholars critical of realism turned to his account of the Peloponnesian War to expose the weaknesses of modern realist— especially neorealist—assumptions and arguments.[12] My own contribution to this literature criticizes realists for lifting lapidary quotes out of context in support of arguments that are not supported by the text as a whole. I offer more nuanced interpretations of his understanding of the causes of the war—having m re to do with Spartan identity than Athenian military power—and of the Melian Dialogue—intended as a critique, not a vindication, of power politics. I further argue that Thucydides might properly be considered the father of constructivism, given his emphasis on the importance of language and its ability in tandem with deeds to sustain or destroy civilization.[13]

The growing interest in mid twentieth century writings, including those of Hans Morgenthau, John Herz, and E. H. Carr, is a natural development as they are the foundational texts of modern realism.[14] They are also appealing because of their

---

[10]Alexander E Wendt, "Anarchy is What States Make of It: The Social Construction of Power Politics" *International Organization,* 46: 2 (1992), 391–425.

[11]John J. Mearsheimer, "E.H. Carr vs. Idealism: The Battle Rages On", *International Relations,* 19 (2005), 139–52; John J. Mearsheimer, Paul Rogers, Richard Little, Christopher Hill, Chris Brown, and Ken Booth, "Roundtable: The Battle Rages On", *International Relations,* 19 (2005), 337–60.

[12]Daniel Garst, Thucydides and Neorealism, *International Studies Quarterly,* 33: 1 (1989), 469–97; Clifford Orwin, *The Humanity of Thucydides* (Princeton, NJ: Princeton University Press, 1994); Paul A. Rahe, Thucydides Critique of Realpolitik, in Benjamin Franke *Roots of Realism* (Portland, OR: Frank Cass, 1996), 105–41.

[13]Richard Ned Lebow, "Thucydides the Constructivist", *American Political Science Review,* 95 (2001), 547–60.

[14]E.H. Carr, *The Twenty Crisis: An Introduction to the Study of International Relations* (London: Macmillan, 1951); John Herz, *Political Realism and Political Idealism: A Study in Theories and Realities* (Chicago, IL: University of Chicago Press, 1951); and *International Politics in the Nuclear Age* (New York: Columbia University Press, 1959).

perspectives on IR and political science. Neorealism claimed to be a scientific theory at the system level, with testable propositions deduced from its central assumptions. Critics were able to demonstrate that it is impossible to restrict the study of IR to the system level that Waltz's principal proposition that anarchy must produce a war-prone, self-help system did not necessarily follow. For someone who claimed that the distinguishing feature of his theory was its scientific rigor, it did not help that Waltz s key conceptual terms of anarchy, power, and polarity were not formulated in a manner that allowed their unproblematic measurement or falsification. In contrast to neorealism, mid-century foundational texts are interested in foreign policy as much as IR. They are steeped in history, acknowledge the importance of the idiosyncratic as w ell as the general, emphasize agents along with structures, accept the limits, if not the impossibility of prediction, and recognize the often determining influence of domestic regimes and politics on foreign policies. In contrast to Waltz, Carr, Morgenthau, and Herz all distinguish material capabilities from power and power from influence. They recognize influence as very much situation dependent, and often related to the ethical basis of the policies in question. Most fundamentally, these realists understand that their theories are products of the epoch and culture that produced them. For all of these reasons, the revisiting of foundational texts is an obvious strategy, not only for scholars who w-ant to attack latter-day realists, but for those who want to use these texts as starting points for reformulations of realism more appropriate to contemporary circumstances and intellectual sensitivities. Several of the scholars who have contributed to this volume are committed to this goal.

The end of the Cold War is not the first major international transformation or upheaval to provoke a return to older texts with the goal of the rethinking of contemporary ideas and approaches and developing alternatives to them. In Roman times, Vergil and Ovid invoked this strategy. Machiavelli, and Hobbes did so in early modern Europe. Machiavelli wrote commentaries on the *Discourses on the First Ten Books of Titus Livius* to resurrect the concept of civic virtue as an antidote to the disorder of Italian city states, and Hobbes translated Thucydides into English in 1651 in response to the English civil war. To put this process in a broader perspective, it is useful to identify and describe the several reasons why scholars are drawn to older texts. I believe there are four such reasons, and they often, if not usually, occur in the order in which I present them.

Justification: Every new text or discourse looks for justification for its arguments or claims. One effective way to do this is to situate ones text or discourse in a tradition that is considered legitimate and respected by the intended audience. Contenders for power in Rome traced their ancestry back Aeneas, if not the gods. The gospels of Luke and Matthew buttress their claims that Jesus is the messiah by tracing his descent back through different routes to King David and through him to Adam. Political philosophers have played this game for millennia, attaching themselves and their works to particular traditions. So do IR theorists. Realists claim descent from Thucydides, Kautilya, Machiaveili, and Hobbes; liberals from

Smith and Kant; Marxists from the eponymous Marx; and constructivists root their enterprise in Nietzsche and twentieth-century philosophy and anthropology. Hans Morgenthau contends that his approach to IR is based on the timeless wisdom of classical Greek and Indian writers. Lesser claims he supports with quotes from more recent authors, leading Barrington Moore in his review of *Politics Among Nations* to complain about his proclivity of "substituting an apt quotation—preferably from an author dead at least a hundred years—for rigorous proof".[15]

Delegitimization: If a good genealogy lends credence to texts and discourses, one way to attack and discredit them is to challenge their descent. This strategy is widely practised in contemporary scholarship. Critics of laissez-faire capitalism have gone back to the writings of Adam Smith, especially his *Theory of Moral Sentiments,* to show that he was concerned about the social consequences of unrestrained capitalism. Critics of crude neopositivism have offered alternative readings of Max Weber to offset demonstrably biased representations of his views about social science by his first English translators.[16] More recently, they have focused attention on the Vienna School to delegitimize the epistemological foundations of King, Keohane, and Verba, *Designing Social Inquiry.* They have pointed out that the w ritings of Karl Popper on which King, Keohane, and Verba rely, were subsequently disavowed by Popper as both unrealistic and unnecessary.[17]

Critics of Morgenthau have engaged in a variant of this strategy. Instead of challenging his reading of the texts he cites in support of his theory, they have contested his readings of waiters whom he wishes to discredit and play off against. They have shown how his characterization of Hersch Lauterpacht, Solly Zuckerman, and other international lawyers of the interwar years as 'idealists' is misleading but in keeping with his goal of setting them up as straw men.[18] Another variant is to go after a text or discourse with works they neither cite nor claim genealogy from but can nevertheless be used to discredit them. Rousseau and Nietzsche have both been mobilized to attack liberalism and neopositivism. Rengger does this with Morgenthau, by foregrounding Michael Oakeshott's critique of his reliance on tragedy as an organizing framework. Brown rightly observes that the current appeal of Carl Schmitt seems to lie in the weapon his writings offer to left-wing critics of liberalism.

---

[15]Barrington Moore, Jr., "Review of Politics Among Nations", *American Sociological Review,* 14 (April 1949), 326.

[16]The first attempt to do this took the form of a biography with extensive excerpts from Weber's work. Reinhard Bendix, *Max Weber: An Intellectual Portrait* (Berkeley, CA: University of California Press, 1960).

[17]Richard Ned Lebow and Mark Lichbach (eds.), *Theory and Evidence in Comparative Politics and International Relations* (New York: Palgrave-Macmillan, 2007).

[18]Brian C, Schmidt, Anarchy, World Politics and the Birth of a Discipline: American International Relations, Pluralist Theory and the Myth of Interwar Idealism, *International Relations,* 16: I (2002), 9–31; Lucian M. Ashworth, "Did the Realist-Idealist Great Debate Really Happen? A Revisionist History of International Relations", *International Relations,* 16, 1 (2002), 33–51.

Reclaiming Texts: Authors go out of style by virtue of their own views, how they have been interpreted by others and the subsequent discrediting of their projects. Friedrich Nietzsche went into decline in the West after he was appropriated by the Nazis in the 1930s, with the active support of his sister, a Nazi collaborator. In the 1960s, Walter Kaufmann wrote an intellectual biography of Nietzsche and translated many of his works to rehabilitate him. He showed that his views, especially about race, had been grossly distorted by Nazi propagandists.[19] Subsequent studies of Nietzsche have carried this project along and by the 1990s made him sufficiently respected for constructivists to find it in their interest to cite him as a forbear.

This strategy of reclaiming texts represents a combination of the first two strategies. The goal is to root one's text or preferred discourse m a text or texts that have traditionally served to legitimize opposing texts or discourses. The extraordinary attention devoted to the newly restored and translated gospel of Judas is motivated at least in part by the interest of many Western Christians to challenge the orthodox interpretation of Jesus and Christianity imposed by the Council of Nicea in 325 C.E. when it was enshrined by canonizing the gospels of Matthew, Mark, Luke, and John. Political philosophers have challenged Catholic readings of Aristotle in the writings of Thomas Aquinas, and reinterpreted his works to serve as foundations for their projects. Lang reminds us that Morgenthau attempted something similar with Aristotle, attempting to use him as a foundation for his claims about the relevance of ethics to foreign policy. Jütersonke might be described as doing the same with Morgenthau. He shows how engaged he was while still in Europe with contemporary debates over the meaning of law, the role norms and the relationship of law to foreign policy. By using Morgenthau to access these issues, he is able not only to enrich our understanding of Morgenthau, but to situate current disputes on these questions in a meaningful historical perspective.

To the extent that the field of IR has a foundational text, it is Thucydides, and it is hardly surprising that he has become the focus of scholars from competing paradigms. In *Tragic Vision of Politics*: I previously noted, one of my goals was to show that Thucydides could rightfully be claimed as a foundational text by constructivists.[20] Much of the recent interest in Morgenthau is by realists who want to use him and other mid-century realists as a jumping off point for the reformulation of this paradigm. Such an effort involves not only a close reading of these texts, but emphasizing aspects of them that were neglected, misread, or simply read differently by earlier scholars.

Texts *as 'Inspirations'*: Close, hermeneutical readings of texts attempt to enter into a dialogue with their authors to recreate as far as possible their understanding of their project and its meaning. Efforts to delegitimize other readings of these texts require that the readings one offers be more credible. The same is true of efforts to root one's own text or preferred discourse in a respected classic. Another, looser kind of reading is possible, and can serve an entirely different purpose. From

---

[19]Walter Kaufmann, *Tragedy and Philosophy* (Princeton, NJ: Princeton University Press, 1968).
[20]Lebow, *The Tragic Vision of Politics*, Chaps. 2–3.

Schelling to Hegel, Hölderlin, Nietzsche, and Heidegger, several generations of German writers and philosophers engaged in 'dialogues' with ancient Greek playwrights in their search for a discourse appropriate to their epoch.[21] Some of these readings use these texts as something akin to Rorschach inkblots on which to project their own yearnings. Hegel's reading of *Antigone*—which stresses the different ethical positions of man and woman, and how events in the tragedy unfold to reveal the need of the spirit to recognize itself in its radical individuality—offers a novel interpretation of the play that tells us a lot more about Hegel than it does about Sophocles.[22] The same is true of Nietzsche's projection of Enlightenment individualism on to aristocratic heroes, or his contention that *Antigone* and *Oedipus at Colon us* are at their core struggles between the sexes and the Apollonian and Dionysian.[23] Greek tragedy was a catalyst for Nietzsche's imagination and led him to ideas that he subsequently read back into texts. He used his interpretations to make his insights and concepts resonate more effectively with his audience. Sigmund Freud's reading of *Oedipus Tyrannus* is even more inattentive to textual detail and historical context, but nobody denies the psychoanalytic utility of the 'Oedipal complex' on the grounds that Oedipus himself clearly did not have such a complex.[24] It would not be productive to evaluate the interpretations of Hegel, Nietzsche, or Freud in terms of the tragedies they wrote about. The more appropriate yardstick is for their originality and philosophical, literary, or medical utility.

What should we make of all of this? We might relish the irony that some of the latest efforts at originality in our field are being played out m a game that has existed for millennia. This iteration differs only in its scale and diligence, both a reflection of the large number of young scholars with political and theoretical agendas, keen to make names for themselves in the profession and for whom texts are the most obvious intellectual resource. Upon further reflection, we might acknowledge that in the social sciences and the humanities this is not only inevitable, but perhaps, a necessary strategy Arguments depend as much on their wrapping as they do on their substance, and the purpose of all arguments is to convince. This is especially true of novel arguments, which must exploit ambiguities in existing discourses to make a wedge for themselves and to be heard. As we do with presents, let us carefully unwrap these arguments, and take pleasure in what we find, and if not, hope that they can be returned, or rewrapped and passed along

---

[21]Dennis J. Schmidt, On Germans and Other Greeks: Tragedy and Ethical Life (Bloomington, IN: University of Indiana Press, 2001), 192–3.

[22]Georg Wilhelm Friedrich Hegel, *The Phenomenology of Mind*, trans. by 1. B. Baillie, ed. George Lichtheim (New York: Harper & Row, 1967), paras. 457, 463–6.

[23]Friedrich Nietzsche, *The Birth of Tragedy*, trans. by Walter Kaufmann (Mineola, NY: Dover, 1995), Sects. 1 and 3.

[24]Sigmund Freud, *The Interpretation of Dreams*, trans. and ed. James Strachey, 3rd edn. (New York: Basic Books, 1955), 'Dostoevsky and Parricide!' in James E. Strachey (ed.), *The Standard Edition of the Complete Psychological Works of Sigmund Freud* (London: Hogarth Press, 1961), vol. 21, 138; and A General Introduction to Psychoanalysis, trans. Joan Riviere (New York: Liverwright, 1935), 29, for Freud's evolving understanding of the Oedipal complex and the play.

to someone else for whom they are better suited. Let us also pay attention to the wrapping paper because it can be as important as the present to the extent that it offers us new insight into old texts approached and read with present-day sensitivities, interests, and understandings.

5.3   Morgenthau and the Post-cold War World

Every generation reads older texts from the perspective of contemporary concerns and with the benefit of knowing Why previous generations turned to—or away—from these texts, what they wrote about them and how they used them to advance their projects. New readings often yield new insights. Recent books and articles about Morgenthau have lived up to this expectation.[25] They explore a set of questions that were previously neglected or considered peripheral. These works emphasize the links between his German and American writings, the debt that both owed to Weber and Nietzsche and the ways his thinking about IR evolved from the deep pessimism expressed in his early post-war writings to the more cautious optimism expressed in his later works. Considerable controversy surrounds his intellectual lineage, especially his degree of indebtedness to Carl Schmitt, the relationship of his legal to his political writings, the extent to which the understandings of politics on which he based his theory of IR were formulated before or after he emigrated from Germany, and the extent of and reasons for his advocacy of a form of transformational politics late in his career The chapters in this volume shed new light on most of these controversies.

Let us return to the problem of transformation, made visibly problematic by the end of the Cold War Neorealists and other realists too, differentiate systems on the basis of their polarity System change occurs when the number of poles changes. Realists associate such with hegemonic wars, brought on, according to power transition theories, by shifts in the balance of material capabilities. Rising powers

---

[25]Andreas Sollner, 'German Conservatism in America: Morgenthau's Political Realism', *Telos* 72 (1987), 161–77; Greg Russell, *Hans Morgenthau and the Ethics of American Statecraft* (Baton Rouge, LA: Louisiana State University Press, 1990); Joel Rosenthal, *Righteous Realists: Political Realism, Responsible Power and American Culture in the Nuclear Age* (Baton Rouge, LA: Louisiana State University Press, 1991); Christoph Fret, *Hans J. Morgenthau: An Intellectual Biography* (Baton Rouge, LA: Louisiana State University Press (1994, 2001); Tarak Barkawi, 'Strategy as Vocation: Weber, Morgenthau and Modern Strategic Studies', *Review of International Studies*, 24: 2 (1998), 159–4; Jan Wilhelm Honig, 'Totalitarianism and Realism: Hans Morgenthau's German Years', *Security Studies*, 5: 2 (1995–96), 283–313; Martti Koskenniemi, The Life title *Civilizer of Nations: The Rise and Fall of International Law 1870–1960* (Cambridge: Cambridge University Press, 2002); Benjamin Mollov, *Power and Transcendence: Hans J. Morgenthau and the Jewish Experience* (Lanham, MD: Lexington, 2002); Campbell Craig, *Glimmer of a New Leviathan: Total War in the Realism of Niebuhr, Morgenthau and Waltz* (New York: Columbia University Press, 2003); Anthony F. Lang, Jr. (ed.), *Political Theory and International Affairs: Hans J. Morgenthau on Aristotle's Politics* (Westport, CT: Praeger, 2005).

may go to war to remake the system in their interests, or status quo powers to forestall such change. For some realists, this cycle is timeless and independent of technology and learning. Others believe that nuclear weapons have revolutionized international relations by making war too destructive to be rational. In their view, this accounts for the otherwise anomalous peaceful transformation from bi- to multipolarity at the end of the Cold War.[26]

The end of the Cold War and disappearance of the Soviet Union, which arc responsible for the ingoing transformation of the international system, are hard to square with these understandings. They were not the result of war. Equally anomalous is that neither Mikhail Gorbachev nor Boris Yeltsin, the Soviet leaders responsible for the changes and concessions that led to the end of the Cold War and the demise of the Soviet Union, were motivated by commitments to preserve or expand the power of the Soviet Union. Gorbachev did not foresee the consequences of his reforms, but Yelstin was fully aware that the Russian Republic's proclamation of independence would hasten the breakup of the Soviet Union.[27] The nature of the system transformation is also the subject of dispute among realists. Waltz insisted that the world remained bipolar in the immediate aftermath of the Cold War.[28] More realists describe it as multipolar, and some prominent American realists contend that it is unipolar. These disagreements reveal the essential ambiguity of the concept of polarity. There are also grounds for questioning its utility as kcy realist predictions based on the assumption of uni- or multipolarity have not come to pass. NATO has prospered instead of collapsing, and allies and third parties alike have looked for ways to ignore, finesse, or resist various American initiatives, but none of them givc evidence of attempting to balance against it.[29]

For classical realists like Morgenthau, transformation is a broader concept and one they associate with shifts in identities, discourses, and conceptions of security. Morgenthau conceives of political systems in terms of their principles of order, and is interested in the ways in which these principles help shape the identities of actors and the framing of their interests. As Richard Little shows, in his view, the success of the balance of power for the better part of two centuries was less a function of the

---

[26]Waltz, 'Theory of International Politics, and The Emerging Structure of International Politics'; Mearsheimer, 'Back to the Future'; William C. Wohlforth, 'Realism and the End of the Cold War', *International Security*, 19 (1994–95), 91–129; Kenneth A. Oye, 'Explaining the End of the Cold War: Morphological and Behavioral Adaptations to the Nuclear Peace?', in Richard Ned Lebow and Thomas Risse-Kappen (eds.), *International Relations Theory and the End of the Cold War* (New York: Columbia University Press, 1995), 57–84.

[27]Robert D. English, *Russia and the Idea of the West: Gorbachev, Intellectuals, and the End of the Cold War* (New York: Columbia University Press, 2000); Archie Brown, *The Gorbachev Factor* (Oxford: Oxford University Press, 1996); Richard K. Herrmann and Richard Ned Lebow (eds.). *Ending the Cold War* (New York: Palgrave-Macmillan, 2004); George Breslauer, *Gorbachev and Yeltsin as Leaders* (New York: Cambridge University Press, 2002).

[28]Waltz, 'The Emerging Structure of International Politics'.

[29]For a review of this literature and the fuzziness of the concepts on which these debates are based, see Richard Ned Lebow, *Rethinking International Relations Theory* (Oxford: Oxford University Press, forthcoming).

distribution of capabilities than it was of the existence and strength of international society that bound together the most important actors in the system. When that society broke down, as it did from the first partition of Poland through the Napoleonic Wars, the balance of power no longer functioned to preserve the peace or existence of the members of the system.[30] International society was even weaker in the twentieth century, and its decline was an underlying cause of both world wars, Morgenthau worried that its continuing absence in the immediate post-war period had removed all constraints on superpower competition. From the vantage point of the twenty-first century, Morgenthau reads as much like a proto-constructivist as he does a realist. For classical realists like Thucydides and Morgenthau, changes in identities and interests are often the result of underlying processes like modernization, and hegemonic war is more often a consequence than a cause of such a transformation.[31] This different understanding of cause and effect has important implications for the kinds of strategies classical realists envisage as efficacious in maintaining or restoring order. They put more weight on values and ideas than they do on power.

Nick Rengger stresses that Morgenthau s understanding of modernization in *Scientific Man versus Power Politics* emphasizes its negative features. In Morgenthau's judgment, the Enlightenment promoted a misplaced faith in reason that undermined the values and norms that had restrained individual and state behavior. In making this association, Morgenthau drew directly on Hegel and Freud. Hegel warned of the dangers of homogenization of society arising from equality and universal participation in society. It would sunder traditional communities and individual ties to them without providing an alternative source of identity.[32] Hegel wrote on the eve of the Industrial Revolution and did not envisage the modern industrial state with its large bureaucracies and modern means of communication. These developments, Morgenthau argued, allowed the power of the state to feed on itself through a process of psychological transference that made it the most exalted object of loyalty. Libidinal impulses, repressed by the society, were mobilized by the state for its own ends. By transferring these impulses to the nation, citizens achieved vicarious satisfaction of aspirations they otherwise could not attain or had to repress. Elimination of the Kulaks, forced collectivization, Stalin's purges, the Second World War, and the Holocaust were all expressions of the transference of private impulses

---

[30]Hans J. Morgenthau, *Politics Among Nations; the Struggle for Power and Peace*, 3rd edn. (New York: Knopf, I960), 160–6; *In Defense of the National Interest*, 60, Paul W. Schroeder, A.J. P. Taylor' s 'International System', *International History Review*, 23 (2001), 3–27, makes the same point.

[31]Lebow, *Tragic Vision of Politics*, Chap. 7 for further development of this argument.

[32]G.W.F. Hegel, *Phenomenology of Spirit* (1307) and *Philosophy of Right* (1821). Charles Taylor, *Hegel* (Cambridge: Cambridge University Press, 1973), 403–21.

onto the state and the absence of any limits on the state's exercise of power.[33] Writing in the aftermath of the great upheavals of the first half of the twentieth century, Morgenthau recognized that communal identity was far from an unalloyed blessing: it allowed people to fulfil their potential as human beings, but also risked turning them into 'social men' like Eichmann who lose their humanity in the course of implementing the directives of the state.[34]

For Morgenthau, the absence of external constraints on state power was *the* defining characteristic of international politics at mid-century. The old normative order was in ruins and too feeble to restrain great powers.[35] Against this background, the Soviet Union and the United States were locked into an escalating conflict, made more ominous by the unrivalled destructive potential of nuclear weapons. The principal threat to peace was, however, political. Moscow and Washington were 'Imbued with the crusading spirit of the new moral force of nationalistic universalism, and confronted each other with 'inflexible opposition'.[36] The balance of power was a feeble instrument in these circumstances, and deterrence w-as more likely to exacerbate tensions than to alleviate them. Bipolarity could help to preserve the peace by reducing uncertainty—or push the superpowers towards war because of the putative advantage of launching a first strike. Restraint was needed more than anything else, and Morgenthau worried that neither superpower had leaders with the requisite moral courage to resist mounting pressures to engage in risky and confrontational foreign policies.[37]

Realism in the context of the Cold War was a plea for statesmen—above all, American and Soviet leaders—to recognize the need to coexist in a World of opposing interests and conflict. Their security could never be guaranteed, only approximated through a fragile balance of power and mutual compromises that might resolve, or at least defuse, the arms race and the escalatory potential of the various regional conflicts in which they had become entangled. Morgenthau insisted that restraint and partial accommodation were the most practical short-term strategies for preserving the peace. A more enduring solution to the problem of war required a fundamental transformation of the international system that made it more like well-ordered domestic societies.

For Morgenthau, the universality of the power drive meant that the balance of power was 'a general social phenomenon to be found on all levels of social

---

[33]H.J. Morgenthau, *Politics Among Nations*, 169. The psychological component of this analysis relied heavily on the earlier work of Morgenthau's Chicago colleague, Harold Lasswell, *World Politics and Personal Insecurity* (New York: McGraw-Hill, 1935), Morgenthau also drew on Hegel.

[34]Hannah Arendt, *Eichmann in Jerusalem: A Report on the Banality of Evil* (New York: Viking, 1964).

[35]Hans J. Morgenthau, *The Decline of Democratic Politics* (Chicago, IL: University of Chicago Press, 1958), 60.

[36]Hans J. Morgenthau, *Politics Among Nations*, 1st edn. (New York: Knopf, 1948), 430.

[37]Morgenthau, *Politics Among Nations*, 1st edn., 169; *Decline of Democratic Politics*, SO; Letter to the *New York Times*, 19 June 1969.

interaction.[38] Individuals, groups, and states inevitably combined to protect themselves from predators. At the international level, the balance of power had contradictory implications for peace. It might deter war if status quo powers outgunned imperialist challengers and demonstrated their resolve to go to war in defense of the status quo. But balancing could also intensify tensions and make war more likely because of the impossibility of assessing with any certainty the motives, capability, and resolve of other states. Leaders understandably aim to achieve a margin of safety, and when multiple states or opposing alliances act this way, they ratchet up international tensions. In this situation, rising powers might be tempted to go to war when they think they have an advantage, and status quo powers to launch preventive wars against rising challengers. Even when the balance of power failed to prevent war, Morgenthau reasoned, it might still limit its consequences and preserve the existence of states, small and large, that constitute the political system. He credited it with having served these ends for much of the eighteenth and nineteenth centuries.

Morgenthau wrote in the aftermath of destructive wars that undermined the communities and conventions that had previously sustained order at home and abroad. He did not think it feasible to restore the old way of life, aspects of which had become highly problematic even before the onset of war. Nor, Craig Campbell tells us, was he attracted to liberal, international projects that aspired to impose limits on state sovereignty. He searched instead for some combination of the old and the new that could accommodate the benefits of modernity white limiting its destructive potential. By 1962, the emigre scholar who twenty years earlier had considered the aspirations of internationalists dangerous would now insist that the wellbeing of the human race required 'a principle of political organization transcending the nation-state.'[39] By the 1970s, he had become more optimistic about the prospects for peace. In his view, detente, explicit recognition of the territorial status quo in Europe, a corresponding decline in ideological confrontation, the emergence of Japan, China, West Germany as possible third forces, and the effects of Vietnam on American power had made both superpowers more cautious and tolerant of the status quo.[40] Of equal importance, their daily contacts, negotiations, and occasional agreements had gone some way towards normalizing their relations and creating the basis for a renewed sense of international community.

Morgenthau's belief in the need for some form of supranational authority also deepened in the 1970s. Beyond the threat of nuclear holocaust, humanity was also threatened by the population explosion, world hunger, and environmental degradation. He had no faith in the ability of nation states to ameliorate any of these problems.[41] If leaders and peoples were so zealous about safeguarding their

---

[38]Morgenthau, *Decline of Democratic Politics*, 49, 81.

[39]Ibid. 75–6.

[40]Morgenthau, *Politics Among Nation*, 5th edn. (1972), preface. Pages 355–6 still reflect the pessimism of earlier editions.

[41]Kenneth W, Thompson, 'Introduction', *In Defense of the National Interest* (Lanham, MD: University Press of America, 1982), v; personal communications with Hans Morgenthau.

sovereignty, there was little hope of moving them towards acceptance of a new order Progress would only occur when enough national leaders became convinced that it was in their respective national interests. The series of steps Europeans had taken towards integration illustrated the apparent paradox that 'what is historically conditioned in the idea of the national interest can be overcome only through the promotion in concert of the national interest of a number of nations.'[42] Paradoxically, if slyly, he envisaged realism, with its emphasis on state interests, as a means of ultimately transcending the nation state.

In the best tradition of the Greeks, Lang observes, Morgenthau aspired to develop a framework that actors can use to work their way through contemporary problems. He insisted that 'All lasting contributions to political science, from Plato, Aristotle, and Augustine to the *Federalist*, Marx and Calhoun, have been responses to challenges arising from political reality. They have not been self-sufficient theoretical developments pursuing theoretical concerns for their own sake.[43] Great political thinkers confronted problems that could not be solved with the tools on hand, and developed new ways of thinking to use past experience to illuminate the present. Beyond this, Morgenthau sought to stimulate the kind of reflection that leads to wisdom and with it, appreciation of the need for self-restraint, especially on the part of great powers. He remains a man for all seasons.

5.4   Texts as Resources

Almost contemporaneous with the invention of writing, political authorities recognized the importance of texts and their need to control them. The Old Testament was codified with this end in mind, and for over a millennia the Roman Catholic church forbade translations of the Bible into the vernacular. At its core, the Enlightenment was an attempt to use reason to destroy tradition and free the individual, and its proponents envisaged texts as powerful weapons in this struggle. Christian and Muslim fundamentalists offer striking contemporary examples of how radical discourses and the interpretations of texts they enable can be used to mobilize political support against the religious and political establishment.[44] Their readings certainly treat the Bible and Koran as Rorschach inkblots, but their

---

[42]Morgenthau, *Decline of Democratic Politics*, 93.

[43]Hans J. Morgenthau, 'The Purpose of Political Science', in James C. Charlesworth (ed.), *A Design for Political Science: Scope, Objectives and Methods* (Philadelphia, PA: American Academy of Political and Social Science, 1966), 77.

[44]Susan Friend Harding, *The Book of Jerry Palwell: Fundamentalist Language and Politics* (Princeton, NJ: Princeton University Press, 2000), shows how fundamentalists use the Bible as a generative text to create new cultural forms. They invoke the Holy Spirit as a unifying interpretive convention that allows ongoing creation. Fundamentalist language is therefore the opposite of a sceptic's literalist reading that searches for contradictions: rather it seeks to integrate, reconcile, and generate, to create hybrid cultural forms rather than separatist ones.

inventive readings of these texts have not prevented them from gaining numerous adherents.

Any examination of the relationship between texts and politics must differentiate between authoritarian and open societies. Texts play a somewhat different role in both kinds of societies, are controlled much more closely in the former and accordingly require different strategies for reinterpreting or discrediting them. Successful decentering of texts in authoritarian societies is also likely to have more immediate and far-reaching implications.

The Soviet Union provides a good illustration of how texts function in authoritarian societies. The writings of Marx and Engels, and then of Stalin became required reading. Stalin's *Short Course of the History of the All-Russian Communist Party*—known everywhere as the *Short Course*—was published in 193S and considered the encyclopedia of Marxism. In 1956, when Khrushchev denounced Stalin's 'Cult of the Personality' at the Twentieth Party Congress, over 42 million copies of the book had been printed and distributed in 67 languages. Intellectuals and writers often had to make ritual genuflections to this work and the Marxist-Leninist canon more generally to have any chance of getting their own works published. Works of fiction had to adhere to the party line, and even when they did, their authors were still at risk when that line changed. In periods of thaw, some experimentation becomes possible, and artists, journalists, and publishers usually pushed against the limits of expression, judging from the Soviet and Chinese experiences, thaws are frequently followed by periods of renewed repression. In this environment, producers and consumers of texts resort to two quite different strategies. They rely on illegal means of communication, which include *samizdat* (mimeo texts passed from hand to hand), wall posters hung furtively at night, underground performances, foreign radio broadcasts, and uncontrolled Internet websites and chat rooms. They also rely on 'double discourse', which uses plot lines and dialogue with hidden references or double meanings, unrecognized or tolerated by the authorities, to satirize the regime and its policies. This is a time-honored practice that has occasionally resulted in great works of literature, philosophy, or art, such as Montesquieu's *Persian Letters* and Kasimir Malevich's painting *Red Cavalry*.

Authoritarian regimes on the wane, especially ones whose ruling elites have lost faith in their ideology, find it increasingly difficult to maintain control over texts. Periods of thaw and repression become more frequent, with overall, if slow, movement towards greater freedom of expression. This has been the pattern m China. A period of thaw can also spin out of control, as in Gorbachev's Soviet Union, leaving leaders little choice but to make whatever accommodation they can with a fast-changing reality Regime survival in these circumstances is difficult because long-suppressed antagonism towards the regime and between different components of the society (i.e., classes, ethnic, and religious groups) can become pronounced.

In more open societies, texts also play an important role, and there are likely to be multiple texts sustaining multiple discourses. Control over them is much less certain, especially if it is exercised by government, and criticism and reformulation of texts is more open. Discourses shift more quickly without disrupting the political order, and their implications for the society, while ultimately profound, are not necessarily evident at the time. Indeed, the reasons why interpretations change, the ways in which they affect the appeal of discourses, and the routes by which changing discourses influence politics remain mysterious, under theorized, and relatively unexplored empirically.

The academic community is all about texts. Reputations are made from pro ducing, interpreting, and critiquing them. Unless we view' this activity as an elaborate *jeu d'esprit*, like games of chess played for their own sake or the reputations and other benefits they confer on those who excel, the production and reading of texts must have substantive implications for the real world. Most academics believe they do influence society, and I know from conversations with colleagues that this belief is especially widespread among scholars of IR, We can all point to books about IR that we believe have had significant impact on foreign policy Morgenthau's *Politic Among Nations* (1948) would probably head the list, E. H. Carr's *Twenty Years Crisis* (1939) might come next. If we relax our criteria to include books not specifically in the field of IR we might add, among others, Carl von Clausewitz, *On War* (1832–37), Alfred Thayer Mahan, *The Influence of Sea Power Upon History* (1890), Houston Stewart Chamberlain, *The Foundations of the Nineteenth Century* (1907), V. fc Lenin, *On Imperialism* (1916), John Maynard Keynes, *The Economic Consequences of the Peace* (1919) and Bernard Brodie, *The Absolute Weapon* (1946). The most recent contender for this august list is arguably, Samuel Huntington's *Clash of Civilizations* (1996), I use the words 'might' and 'arguably' because there is no rigorous way to determine the influence of books on the thinking of people in general, on leaders in particular and on the policies that either the public or leaders support or adopt. Sales of books are at best an imperfect indicator of influence. Many non-fiction books that make best-seller lists may be purchased but not read, as seems to be the case most recently with Bill Clinton's memoirs. Leaders sometimes identify books that had an enormous influence on them; for Field Marshal Moltke the Elder it was Clausewitz, and for George Marshall, Thucydides. There is, of course, no way of knowing it this was really the case, or if they were drawn to these books because they justified policies they already favored.

Assuming for the sake of argument that my list is a reasonable, if by no means an exclusive one, it suggests that scholarship in the most direct sense has only a limited influence on the practice of international relations. Ten plus books in more than a century-and-a-half is not very many. And only one of them was published since 1950 (not counting new editions), and it does not have an unambiguous claim to the list. The three books published at the beginning of the Second World War or in its aftermath had more diff use influence. They helped to shape how a generation of students and influential people thought about the post-war world and problems of security. Thomas Schelling's *Arms and Influence* (1966) might recommend itself to the list for the same reason.

Most of the books fit into two categories: they pertain to war and weapons, or tap and play on primordial fears of their readers. Clausewitz, Mahan, Lenin, and Brodie all fit into the first category, and Chamberlain and Huntington into the second. Clausewitz—misread by Germans and French alike—helped to justify offensive strategies and decisive battles. Mahan provided the logic and justification for America and Germany's naval build-ups in the latter part of the nineteenth century. Lenin's views had a direct influence on Soviet and Chinese foreign policies. Brodie provided the logic and justification for deterrence and mutual assured destruction. Chamberlains book combined social Darwinism with racism, propagated the idea of an Aryan race, and was widely read in Europe. It was the inspiration for Gobineau's equally racist book in France, and had a major impact upon Richard Wagner. Chamberlain was invited to court by Kaiser Wilhelm and was acknowledged by Adolf Hitler as having been instrumental in his racist thinking.

There is a paucity of books on the list that have peace as their principal message or are devoted to strategies for conflict management and resolution. There are many fine studies of this kind, but only one—Keynes, *Economic Consequences of the Peace*—has become a best seller, and its author almost a household name. Keynes's book encouraged widespread disenchantment with reparations in Britain and the US, and was instrumental in bringing about the Dawes Plan (1923) and the Young Plan (1930). Another work on conflict management that achieved a lesser degree of influence in the policymaking community is David Mitrany, *The Road to Security* (1944), which advocated functionalism as a strategy of conflict management. Academic writing on strategic arms control and the relationship between economic development and peace was also important, but there are no one or two books that stand out above all others in this connection.

Another disturbing finding is that the seeming influence of a book bears little relationship to its quality In the first instance, this is due to the market; scholarly books lack commercial appeal. What publishers bring before the public is what they think will sell, and in the non-fiction category, this is most often simple books with simple messages. Huntington is appealing for precisely this reason. The publishing playing field is far from even. Books that praise a country's leaders (when they are popular) and build or support its myths are far more likely to get published and sell than those that are critical. In most countries, media outlets are controlled by large corporations, some of whose leaders have political agendas and support and publicize books advancing their points of view. The success of the neocons is at least in part due to their ready access to conservative magazines, publishing houses, and television networks. Other forums for publicity rarely feature serious works. How often have you seen a professor interviewed by Oprah Winfrey? Reading as a pastime is declining, as witnessed by the shrinking number of quality newspapers and their readerships. The net effect is to marginalize books in general, and serious books in particular.

There are alternative networks for selling books, and these sometimes achieve phenomenal success. The multiple volumes of Rev Tim La Haye's and Jerry Jenkin s *Left Behind Series* have generally grabbed the number one position on the *New York Times* best-seller list their first week of release. As of September 2005, *Harry Potter* books had sold 30 million copies worldwide, while the *Left Behind* series sold twice that number in the United States alone. *Left Behind* peddles a primitive, parochial, and retributive version of morality: good and evil is sharply delineated; those who embrace Jesus go to heaven and everyone else suffers eternal damnation.[45] The books are very much about international politics. Their plots revolve around a conflict between Jesus and the anti-Christ (a Romanian Secretary General of the United Nations), which reaches its climax at the Rattle of Armageddon. The books offer a thinly veiled policy agenda, which is frequently reinforced at church meetings where the *Left Behind* books are discussed and sold. Measured in terms of the number of people they reach in the United States, and increasingly abroad as well, they undoubtedly are having more influence than any book or books produced by any scholar or community of them.

When scholarly or otherwise weighty books achieve influence they generally address newsworthy problems, are published at an opportune moment, and often have prominent sponsors. Brodie's book offered a useful conceptual framework for thinking about nuclear weapons, a problem on the mind of every military officer, foreign policy official, and much of the educated public in the aftermath of Hiroshima and Nagasaki. Brodie and Morgenthau both benefited from the enormous public interest in America's pre-eminent role in the world after the Second World War, the onset of the Cold War, and the search for ways of dealing with the Soviet Union short of destructive war Clausewitz *On War* was neither newsworthy nor published at a good time, but had a very prominent sponsor. The posthumous publication of Clausewitz's works in 1832–37 made little impact until Germany stunning triumphs over Austria in 1866 and France in 1870–71. When Field Marshal Helmuth von Moltke, the architect of those victories, told the press that *On War* had been his campaign bible, Clausewitz became an overnight sensation and his work was translated into a score of European languages.[46]

Even this cursory review of the world of IR scholarship and its public and policy impact suggests a sharp disconnect between the practices expected to govern the world of scholarship and those that are in fact operative in publishing and politics. The norms of scholarship are those of science: data and research should be shared and subject to critique, good scholarship is evaluated in terms of an ever-evolving set of practices, and good scholarship is expected to drive out the bad.

This expectation rests on two rarely articulated and untested underlying assumptions. The first is that the appeal and legitimacy of any scholarly discourse within the academy depends on its intellectual rigor. If so, texts can be expected to lose favor and decline in perceived importance to the degree that they can be shown to be inconsistent, at odds with the evidence or based on weak conceptual foundations. Texts and discourses come and go, but rarely, it seems, on the basis of their demonstrated excellence or lack of it. Neorealism took enough hits to sink the *Bismarck,* but still it stayed afloat, if not always able to proceed at full steam, until the end of the Cold War. External developments finally made it list dangerously and primarily because they foregrounded a new set of problems for theorists to which

---

[45]LaHaye and Jenkins, *Left Behind*, 12 volume series.

[46]Eberhard Kessel, *Moltke* (Stuttgart: K.F. Koehler, 1957), 108.

neorealism did not seem relevant. Marxism confronted all kinds of intellectual difficulties but remained an important paradigm in Western scholarship until the collapse of communism in Eastern Europe and the Soviet Union, regimes that were arguably communist in name only. Deterrence theory remains alive and well despite compelling empirical critiques that immediate deterrence rarely succeeds. Rational choice positively prospers despite its inability to account for the so-called voting paradox, or much of anything else.

The second assumption is that there is some connection between academic and policy discourses. If the former exist in a hermetically sealed world of their own, there is little incentive for someone ultimately interested in policy to engage them. For the most part, I think it fair to say that elite and mass discourses exist on parallel tracks with few cross switches. There are times when they intersect, and with powerful consequences. Thucydides and Plato wrote at such a moment in fifth-century Greece. Tocqueviile describes a similar development on the eve of the French Revolution, which he attributes to the influence of Enlightenment writers and propagandists.[47] Elites are at least as ignorant of mass discourses. My informal survey of academic colleagues at a recent meeting of the International Studies Association revealed that hardly any of them have ever heard of, let alone read, any of the volumes of the *Left Behind* series. Only four Ivy League libraries have any of the volumes.[48]

Elite-academic intercourse is more common. In Europe, the elite press reviews serious books and reports on key academic debates. In Germany, the so-called *Historikerstreit* over the Third Reich and the singularity of the Holocaust received wide coverage.[49] In the United States, there is considerable movement of people between the academe and Washington, but much less traffic in texts and ideas. Few academic books are found in governmental offices or on the night tables of top officials. In the 1980s, every IR scholar and graduate student was familiar with Waltz's *Theory of International Politics*. For better or worse, it was largely unknown to policymakers. To the extent that policymakers and their advisors have any time or inclination to read, they are likely to turn to history and biography, not to political science* Prominent academic historians and presidential biographers like Arthur Schlesinger, Jr., Robert Dallek, and Ron Chernow sell widely, and are not infrequently cited, even quoted, by officials.

The uphill struggle IR scholars face in presenting their research to a wider audience should make us respectful of those few members of our profession who have succeeded Morgenthau, Carr, George Kennan and Raymond Aron all reached beyond the academy, but only Morgenthau was a political scientist. Carr and Kennan were historians, and Aron, a sociologist. Morgenthau had additional

---

[47]Alexis de Tocqueville, *The Old Regime and the French Revolution*, trans. Stuart Gilbert (Garden City, NY: Doubleday, 1955), Part 3, Chaps. 1–2, 138–57.

[48]Borrow Direct! Ivy League Library search, November 2005.

[49]Richard Ned Lebow, Wulf Kansteiner and Claudio Fogu (eds.), *The Politics of Memory in Postwar Europe* (Durham: Duke University Press, 2006) for a more general discussion of the intersection of academic and elite discourses in Europe.

handicaps. He was a foreigner who spoke with a strong accent, wrote theory, not biography or historical narratives, and was generally critical of administration policy, especially during the Vietnam War. Viewed in this light the success he had in finding fora from which he could speak truth to powder is all the more impressive.

5.5 Conclusions

To this point in my argument I have written about the external benefits of revisiting texts, that is about their instrumental ability to influence contemporary theoretical debates and foreign policy decisions. There are internal benefits as well, and I would like to conclude my chapter and our book by directing our attention to them. This provides a more upbeat ending because unlike the external benefits, which while real, are difficult to achieve, the internal benefits are within the grasp of each of us.

Scholarship differs from journalism in its emphasis on theoretical underpinnings, strivings for conceptual rigor and terminological precision, and adherence to sophisticated rules for the selection and evaluation of evidence. Scholars hold themselves to different, some would say, higher standards than journalism, and certainly of the kind of writer that engages in partisan advocacy. On one level, the practice of science, and scholarship more generally, is dependent on an ever-evolving set of protocols that we use to conduct and evaluate research. At a deeper level, science and good scholarship in any discipline or field depend on the ethical commitments of practitioners. These commitments not only make us aspire to play by the rules, they encourage us to improve our ability to conduct good scholarship. And here is where the reading of texts comes into the picture. They teach us by example, by their quality enduring appeal and, in the cases of great texts, the normative commitments that drove their authors and can still motivate us.

Revisiting Texts, especially when they represent different traditions, can teach us humility while widening our intellectual perspectives. We learn that even the greatest works—especially in our field, but in all social science—have serious weaknesses, become rapidly dated in some ways, and incorporate ideas, perspectives, or approaches to problems that have subsequently lost legitimacy. Contributors to this volume tell us that Aristotle is an important figure for Morgenthau, but that his reading of him is at times superficial. They point out that Morgenthau's treatment of the concept of the balance of power is confusing, if not contradictory, that he never reconciles thermonuclear weapons and his resulting support for supranational authority with his fundamental principles of realism, and that from our perspective, his discussion of alliances and balancing appears overly mechanical. We can denigrate Morgenthau and any other respected text of the past for their incompleteness, inconsistencies, occasional superficiality, and even downright errors, or we can recognize that works greater than any we are likely to write have serious flaws. This is a particularly important lesson for graduate students, who are taught from their first day in class to bring to light all the conceptual and empirical shortcomings of everything they read. They are rarely instructed to appreciate the difficulties and limitations under which scholars work, and may develop unrealistically high and possibly crippling expectations about what they might be able to achieve.

Even though most of us know better, it is easy to slip into the mindset that we have some privileged vantage point outside of history that provides us a timeless perspective on politics. Revisiting texts can expose this pretension—particularly prevalent among those IR scholars from quantitative and modelling traditions who characterize what they do as 'normal science'. Revisiting texts, even the greatest of them like Thucydides, Plato, and Aristotle, reveals how culture- and time-bound they are, even if some of their insights appear to have withstood the test of history These works, of necessity, reflect the contemporary interests of their authors, usually embody concepts and categories common to their era as well as modes of expression and language. The older they are, or the further away their culture from ours, the more work we must put in to understand these texts in anyway approaching the intentions of their authors. Not only their framing of problems, but their solutions may strike us as wrong or simply anachronistic. Here too we need to look into the mirror and recognize the extent to which our publications reflect parochial perspectives and forms, and have 'use by' dates of much shorter duration than the classics we so much admire.

Revisiting texts broadens our horizons by demonstrating that no text, no matter how great, has a monopoly on the truth. We learn that the best of texts convey partial truths, and that to develop a more holistic understanding of the politics of any period, or of the subject of politics more generally, we need to read and assimilate the insights of multiple, often competing texts and perspectives. Recognition of this truth should teach us tolerance, respect for people and works in differing traditions, and encourage us to learn from them.

For all of these reasons, Morgenthau, who speaks to us across the abyss of Weimar, the Nazi era, the Second World War, and the Cold War, still has much to teach us.

Chapter 6
Constructing Cause in International Relations

Richard Ned Lebow

> Our idea ... of necessity and causation arises entirely from the uniformity visible in the operations of nature, where similar objects are constantly conjoined together, and the mind is determined by custom to infer the one from the appearance of the other.[1]
>
> David Hume[2]

6.1 Introduction

Social science is the ultimate Enlightenment project. Through reason it aspires to order and understand the social world, and implicitly to reorder it on the basis of this knowledge. The Enlightenment and social science alike were inspired in part by breakthroughs made by scientists in early modern Europe in understanding the physical world. The work of Galileo, Torricelli, and Newton generated the expectation that the physical world might be understood in terms of law-like statements, all of which could be unified in a larger theoretical edifice. Social science—also influenced by socio-political reflection and pre-Darwinist biology—modeled itself on this understanding of physics, which is still reflected in positivism and its various guises.[3]

---

[1]This text was first published as: "*Constructing Cause in International Relations*" (Cambridge: Cambridge University Press, 2014), Chap. 1 (pp. 12–45). ISBN: 9781139699273. The permission to republish this text was granted on 24 June 2015 by Clair Taylor, Senior Publishing Assistant, Legal Services, Cambridge University Press, Cambridge, UK.
[2]Hume, *Enquiry Concerning Human Understanding,* VIII.1.5.
[3]Social science is also the product of eighteenth- and nineteenth-century socio-political reflection that developed independently of the natural sciences. The writings of Vico, Rousseau, and Saint-Simon were especially notable in this regard. Pre-Darwinist biology, and its organicism, vitalism, and proto-functionalism, also contributed to social science. These diverse strands of development conceived of cause in different ways and encouraged diverse perspectives in social science. A standard narrative developed that downplays diversity and emphasizes the scientific component, especially classical mechanics.

© The Author(s) 2016
R.N. Lebow (ed.), *Richard Ned Lebow: Major Texts on Methods and Philosophy of Science*, Pioneers in Arts, Humanities, Science, Engineering, Practice 3,
DOI 10.1007/978-3-319-40027-3_6

The physical sciences have moved away from what was at the outset a religiously inspired vision of the world as a machine set into motion by god its architect. Quantum mechanics and the Copenhagen Interpretation it prompted have brought most physicists around to a view of the world as indeterminate and unpredictable at its core, although subject to robust, often probabilistic generalizations, at the macro level.[4] Investigation of non-linear processes—beginning with Poincare's solution to the three-body problem—also encouraged greater awareness of contingency, confluence, and the limits of prediction.[5] At least some philosophers of science have challenged the feasibility of the longstanding goal of physics to produce equations of universal validity that explain the universe. Nancy Cartwright argues for a 'dappled' view of the world in which different laws contribute to describing events but cannot be assimilated to an overarching, hierarchical frame-work.[6]

In the 1920s, the Vienna Circle conceived of physics as an enterprise in which truth claims could be assessed by logically derived procedures. King, Keohane, and Verba's, *Designing Social Inquiry*—perhaps the most widely used methods text in US graduate political science courses—embraces this epistemology even though it has long been repudiated by its principal authors and other philosophers of science.[7] Typical of behavioral social science, King, Keohane, and Verba are also attracted to a version of Carl Hempel's deductive-nomological model, a response to the inadequacy of constant conjunction. King, Keohane, and Verba are careful to distinguish association from cause, but their unremitting focus on correlation perhaps unwittingly encourages the widespread practice among quantitative researchers of "throwing caution to the wind" when they make policy recommendations that assume causal connections.[8]

Classical mechanics remains the model of science for many social scientists.[9] The social world nevertheless differs in fundamental ways from its physical and biological counterparts. There is nevertheless still something important to be learned from physics. Its fields and even subfields have developed different understandings of cause and its utility. Philosophers of science have devoted much thought to the problem of cause. Here, too, there is no consensus, as numerous frameworks compete with one another, all of them recognizably problematic.

---

[4]By indeterminate physicists I mean that one cannot predict the future state of the system.

[5]Gleick, *Chaos;* Nicholas and Prigogine, *Exploring Complexity;* Bak and Chen, "Self-Organized Criticality"; Byrne, *Complexity and the Social Sciences;* Urry, *Global Complexity;* Jervis, *System Effects.*

[6]Cartwright, *How the Laws of Physics Lie* and *Dappled World.*

[7]King, Keohane, and Verba's, *Designing Social Inquiry,* Chap. 1.

[8]Kinkaid, "Mechanisms, Causal Modeling, and the Limitations of Traditional Multiple Regression".

[9]Salmon, "Four Decades of Scientific Explanation".

In the tradition of Cartesian dualism, analytical philosophers in the 1950s made a sharp distinction between mental life and physical causes. The latter were seen as independent of the former, and many philosophers believed that mental life might be explained by physical, i.e., external causes. Given this orientation, there was no interest in looking at unobservables, like reasons, motives, or thoughts, because they were not considered necessary to account for behavior. Analytical philosophy provided intellectual support for the behavioral revolution in psychology, economics, and political science.

Significantly influenced by the linguistic turn, contemporary analytical philosophy has come to understand reasons, and mental life more generally, as important. The question is no longer their relevance, but how to study the connections between emotion and reflection on the one hand and behavior on the other. This orientation has gradually penetrated the social sciences. Economics has embraced experimentation as a means of exploring this relationship, and psychology, which formerly ignored emotions, or sought to account for them cognitively, has begun to study emotions in their own right. Political science and international relations are moving slowly in the same direction. Recent work, drawing on neuroscience or Greek philosophy, stresses the positive as well as negative role of emotions and properly directs our attention to the conditions associated with these divergent outcomes.[10]

Studies of war nicely illustrate this evolution. Traditional historians looked for underlying reasons of wars, but most gave equal emphasis to agents and their motives. In the postwar era, behavioral approaches became dominant in international relations and led to a fixation on finding regularities or so-called structural determinants of war (e.g. imbalance of power, uneven economic development, absence of institutions able to coordinate). The former approach undergirds the correlates-of-war project, and the latter, balance of power, power transition, and liberal institutionalism. Rationalist approaches to war frame the problem in terms of decisions made by leaders, but deny meaningful agency by explaining those decisions in terms of external constraints and opportunities. My *Why Nations Fight* represents a sharp break with this tradition, as it foregrounds motives as causes in its attempt to explain the origins of war and its frequency.[11]

Constructivists reject all forms of mechanical causation. They insist that behavior cannot be reduced to material causes; it is social all the way down. 'Truth' is not a property of the world, but a product of theoretically informed and evaluated propositions about it. It follows that we never evaluate propositions against the

---

[10]Clore, "Cognitive Phenomenology" and Clore et al., "Affective Feelings as Feedback"; Damasio, *Descartes' Error;* Gray, *Psychology of Fear and Stress;* Marcus, Neuman, and Mackuen, *Affective Intelligence and Political Judgment;* McDermott, "Feeling of Rationality"; see Lebow, "Achilles, Neuroscience and International Relations," for a review.

[11]Lebow, *Why Nations Fight.*

so-called real world, but against 'facts' that are at least in part a creation of our theories. Once we accept that facts and concepts alike are subjective creations, we are forced back on an understanding of the world that shares much in common with the nineteenth-century concept of *Geisteswissenschaft*.[12] The study of international relations moves closer to cultural studies. This is an underlying assumption of my *Cultural Theory of International Relations*, which maintains that different value systems generate distinct logics of cooperation, conflict, and risk taking, and also different kinds of hierarchies and conceptions of justice.[13] These value systems tap universal human motives, but their expression and relative importance are mediated by culture and the understandings people have of their cultures.

The behavioral revolution was based on the premise that the social world resembles the physical world and that knowledge in both consists of regularities and laws that describe them. If the social world is fundamentally different, the search for regularities may be of limited relevance. And in the social world there are few, if any, constant conjunctions. Even such seeming robust regularities like supply and demand and price and demand depend on specific underlying conditions. Increases in demand do not always prompt efforts to provide greater supplies, and the relationship between price and demand is not always inversely related. High-end retailers and restaurateurs have long understood that high prices can stimulate demand.

Hume's formulation of causation, on which regularity theories are based, assumes a world of loose and separate existences in which we look for constant conjunctions. It is embedded in Cartesian dualism and its understanding of mental activities as the product of reason.[14] Social regularities are imperfect because the world is not composed of separate and independent things as Hume supposes. It more closely resembles a web, whose seemingly distinct features are the products of the categories we impose. Any kind of association tells us what *may* happen, not what will happen. At best, associations are starting points for constructing narratives into the past and the future.

Social regularities are generally short-lived because the context that sustains them is always evolving; sooner or later it changes in a way that undermines them. This phenomenon is also attributable to reflexivity, which is another important way in which the social world differs from its physical counterpart. As Max Weber observed, the half-life of any regularity is short because people take it into account once they become aware of it. While a few regularities may be strengthened by

---

[12]Schutz, "Social World and the Theory of Social Action" and "Common-Sense and Scientific Interpretation"; see Searle, *Construction of Social Reality,* on the social construction of concepts and facts.

[13]Lebow, *Cultural Theory of International Relations*.

[14]*Jackson, Conduct of Inquiry in International Relations*, pp. 24–40.

awareness of them, most arguably are not, making social science based on Humean assumptions a self-defeating enterprise—unless one looks at the same time for the features of context that sustain any observable regularities.[15]

For Hume, imagination is responsible for abstract ideas, of which causation is one. The mind receives, organizes, and stores sensory input. Our impressions do not constitute knowledge about the world, but help us cope with it.[16] Hume was a close reader of Newton and subscribed to his belief that rational human inquiry promotes a progressive harmonization and simplification of our understandings of the world, which we treat as "laws of nature."[17] Hume follows Newton in relying on the experimental method, but experience, responsible for the inferences, limits us to observables. We can only guess at underlying causes.

Despite recent and unpersuasive attempts to read Hume as a realist, the great empiricist was clear that causation was not a feature of the world.[18] His approach to it was empirical and reflected what he observed of human practice; people intuitively made causal connections when they observed constant conjunctures and used them to make predictions. Causation for Hume is one of a number of conceptual tools that enable human society.[19]

Stephen Turner has described the desire to make inferences—Humean and other kinds—as "a metaphysical itch that seems not to want to go away."[20] This tendency may be 'hardwired' into us, and, as Hume realized, an effective tool for coping with the world. In recent decades, psychologists have identified a number of cognitive biases that appear to serve similar functions; they help people make rapid decisions about what information to pay attention to and how to interpret it. Research demonstrates that these biases are ubiquitous and sharply at odds with rational procedures for estimating probabilities or making attributions.[21] Psychologists distinguish the former from the latter, and political scientists should do the same when thinking about causation. Humean association captures folk practice. It is an unwarranted leap of faith to assume it appropriate to scientific inquiry.

How then should we think about the concept of cause and its relevance to our scholarly enterprise? In the remainder of this chapter, I review different understandings of cause. I have two intellectual starting points: physics and philosophy of science. I do not model my approach on physics, but turn to this field to make an

---

[15]Lebow, *Forbidden Fruit,* Chap. 1 elaborates on this claim.

[16]Hume, *Enquiry Concerning Human Understanding,* IV.II.14–15, V.1.3–5.

[17]Newton, *Principia,* p. 92.

[18]See Strawson, *Secret Connexion,* for Hume as a realist, and Richman, "Debating the New Hume," for a refutation. See Read and Richman, *NewHume Debate,* for a fuller discussion.

[19]Hume, *Enquiry Concerning Human Understanding,* III.2 and VIII.1.5.

[20]Turner, *Social Theory of Practices,* p. 9.

[21]Kahneman, Slovic, and Tversky, *Judgment under Uncertainty.*

important conceptual point. If physics has no general approach to causation, but rather field and subfield specific ones that have emerged in the course of research, there is no reason to assume that any particular formulation of cause should have precedence in the social sciences, including international relations. Rather, we should follow physics by developing domain-specific understandings that seem useful and appropriate.

My second starting point is philosophy of science, where the nature of cause and its relevance to the physical and social worlds have been most intensively studied. I draw on relevant literature to demonstrate the lack of consensus among philosophers and their recognition of the logical problems and empirical limitations of any understanding of cause. I read this literature to suggest that those of us who engage the social world must nevertheless still invoke some concept of cause to make sense of it. Here, too, however, dissensus suggests that we are free to work with whatever conception we consider germane to our purposes.

6.2 Physics and Cause

Even before the development of quantum mechanics, some physicists questioned the long-held belief in a deterministic universe, on which most understandings of cause rested. In 1909, Franz Exner suggested that all natural events were governed by chance but gave the false appearance of determinism because of the large number of events involved. In the late nineteenth century, Johannes von Kries and Theodor Fechner adopted a Kantian approach to probability. After World War I, they were followed by von Richard von Mises.[22] Moving away from determinism did not require giving up the concept of cause, but it did demand rethinking it and developing understandings commensurate with more probabilistic approaches to science.

In some fields, most notably quantum mechanics, cause became increasingly problematic. Beginning early in the twentieth century, experimental results made it all but impossible to develop causal accounts of some quantum phenomena. Three problems stand out: some events that might otherwise be considered causal are unconnected by any continuous process; some quantum-level processes appear to have instantaneous effects across large distances, and some quantum effects occur before their putative causes.[23] Physicists acknowledge that conventional understandings of causality are also violated whenever an atom emits a photon or a nucleus decays, because none of these events appear to be triggered by a preceding one. Most troubling of all are events for which causal explanations would require

---

[22]Lange, "Causation in Classical Mechanics".

[23]Healey, "Causation in Quantum Mechanics".

instantaneous, that is, superluminal, communication.[24] In the corner of quantum mechanics where these phenomena are observed, physicists have abandoned the concept of cause.

Many other physicists came to consider the universe indeterminate and the concept of cause irrelevant, even a hindrance, to their project.

The 1920s witnessed a reaction to this development. Hans Reichenbach affirmed his belief in causation in the form of law-like function statements of physics.[25] Moritz Schlick—a leading figure in the Vienna School—asserted that everything that happens in nature is subject to valid laws. He subsequently relaxed this belief to accommodate quantum mechanics and the Heisenberg principle and adopted the position that only utter chaos could be described as non-causal. Schlick became an advocate of the regularity view of causation and came to insist that the sole criterion for lawfulness *(Gesetzmässigkeit)* was prediction. In this hollowed-out Humean understanding, causality is nothing more than a directive "to seek regularity" *(Regelmässigkeit).*[26] More recently, J.L. Mackie praises the regularity approach on the grounds that it "has at least the merit that it involves no mysteries, no occult properties like necessity."[27]

In the 1960s, Richard Feynman famously argued that quantum mechanics compels us to reject causal determinism as a condition of nature or goal of science.[28] More recently, Brian Skyrms evaluated seven theories of causation against the empirical findings of quantum phenomena and found them all wanting.[29] Many philosophers have adopted this position, although some physicists remain committed to determinism.[30] The most renowned affirmation of this kind remains the Einstein, Podolsky, and Rosen paper of 1935, which expresses the hope that one day a deterministic theory would emerge that could integrate quantum mechanics and its seeming indeterminacy.[31]

Statistical mechanics is applied to a diverse range of macroscopic objects and their behavior, including thermal and magnetic phenomena and the structure of matter in its four phases (solid, liquid, gas, and plasma). It is a hodgepodge of approaches and formulations, but central to it are the concepts of equilibrium, irreversibility, entropy, and temperature, and the laws governing them. The seemingly bizarre features of quantum mechanics do not feature in statistical mechanics,

---

[24]Shimony, *Search for a Naturalistic World,* p. 151; Dickson, "Non-Relativistic Quantum Mechanics," pp. 390–1.

[25]Reichenbach, "Kausalproblem in der Physik" and "Principle of Causality and the Possibility of its Empirical Confirmation".

[26]Schlick, "Kausalitat in der gegenwartigen Physik"; Stoltzner, "Logical Empiricists".

[27]Mackie, *Cement of the Universe,* p. 60.

[28]Feynman, *Character of Physical Laws,* p. 147.

[29]Skyrms, 'EPR'.

[30]Bohm, "A Suggested Interpretation of the Quantum Theory in Terms of *'Hidden' Variables, I and II" and* Causality and Chance in Modern Physics.

[31]Einstein, Podolsky, and Rosen, "Can Quantum Mechanical Description of Physical Reality Be Considered Complete?".

but physicists in this field, for the most part, do not concern themselves with cause and do not feel the need to do so. They are, however, interested in showing that the laws governing thermodynamics reflect more fundamental features of the world, an ongoing project that has not met with much success.[32]

Statistical mechanics follows a strategy widely practiced in physics and chemistry: devise some description of the transient states of a system and use it to generate law-like statements that predict its future states based on knowledge of the current one. The state of a system can be characterized by the position and momentum of its components. When their statistical distribution is known, the laws of dynamics allow prediction of future states. This operation relies on the artificial assumption that a random probability distribution can be substituted for knowledge about the local conditions in the system. The laws of dynamics offer no explanation for why systems tend to equilibrium or the means by which it is achieved. Physicists invoke non-equilibrium theory to justify the assumption that most systems exist most of the time in or near a state of equilibrium, although they recognize steady-state systems sometimes appear spontaneously, decay, or move to new equilibria.

Another example from statistical mechanics concerns crystallization, which represents a transformation from a local to more general order. When this occurs, each atom or molecule assumes a fixed position relative to the others in the crystal. Something similar happens in ferromagnetism, where the spins of atoms are consistent only with those of adjacent atoms until ferromagnetism creates a uniform alignment throughout the field. There are currently no explanations for these phenomena; they would have to be derived from a robust nonequilibrium theory. In the absence of such a theory, physicists assume the conditions necessary for long-range order and base their predictions on them. Generally known as renormalization group explanations, they depend on formal similarities in the transitions of many different systems from local to general forms of order.[33]

Most systems prepared in non-equilibrium states move to equilibrium. To ask why and how this happens, physicists must deploy some underlying understanding of the statistical nature of things. One candidate is the Boltzmann-Ehrenfest principle of invariant systems; it attempts to account for why most systems spend most of their existence at or near equilibrium. Deviations from it are assumed to be frequent but less frequent as their order of magnitude increases. Small deviations are generally self-correcting, while larger ones have the potential to move systems out of equilibrium.[34]

These transcendental 'explanations' are close to the Kantian conception of cause. To formulate laws of thermodynamics, physicists had to construct abstract, noumenal worlds that incorporated conditions theorized to be fundamental to the universe or its understanding. From them they also derive warrants for inference.

---

[32]Sklar, *Physics and Chance*, p. 5.

[33]Lange, "Causation in Classical Mechanics".

[34]Sklar, *Physics and Chance,* Chaps. 5–7.

The most important is that of equilibrium. In many empirical situations, distributions close to equilibrium are found. Helium gas at room temperature and the earth's atmosphere are cases in point. In non-equilibrium statistical mechanics, physicists recognize that the noumenal worlds they create are highly unrealistic in that they rely on initial conditions that are highly unlikely according to the laws of statistical mechanics. They assume a specific and wildly improbable initial set of conditions for the big bang and also for low-entropy initial states. To account for nonequilibrium behavior they must posit that physically isolated branch systems have entropies that increase simultaneously and in parallel with that of the system as a whole.[35]

A third understanding of cause has developed in cosmology, a largely historical science, like geology and evolutionary biology. Historical sciences are those in which initial conditions are extremely important and critical temporal sequences extend into the past-back to the big bang in the case of cosmology. These sciences rely primarily on observation as the ability to conduct laboratory experiments is very limited. Comparison is accordingly difficult, which makes scientists in these fields alert to the possibility of simulations and natural experiments. A telling example of the latter is the remarkable opportunity provided by Supernova 1987A for observing neutrinos. Japanese and American teams detected neutrino events almost coincident with the discovery of the light from the supernova.[36, 37] These events provided useful information for modeling supernovae and confirmation of theoretical expectations that supernovae would produce intense neutrino fluxes.

Cosmology is like geology and evolutionary biology in that the history of the universe is a one-time run like that of the earth and its life forms. It cannot be compared to other universes, although there is much speculation about whether they exist and what they might be like. Cosmology is also historical in the sense that the creation and early expansion of the universe are critical to explaining the existence and distribution of matter, and other important features of today's universe. When queried about explanation, many physicists say it involves accounting for something by reference to general principles. There is nevertheless a growing recognition in cosmology of the importance of singular causation. In our sample of one universe, we have no way of knowing if its initial conditions and current properties are contingent or expressions of laws, and, if laws, which of them are significant. It is possible that laws of nature are invariant, but that their parameters have assumed different values over the course of our universe's history. It so, then some processes may be responsible for these values, reinforcing the need, Lee Smolin contends, to understand cosmology in an evolutionary context.[37]

---

[35]Ibid.

[36]Arafune and Fukugita, "Physical Implications of the Kamioka Observation of Neutrinos from Supernova 1987A"; Bionta, Blewitt, Bratton et al., "Observation of a Neutrino Burst in Coincidence with Supernova 1987A in the Large Magellanic Cloud".

[37]Smolin, Life of the Cosmos, p. 77.

In all historical sciences there is a commitment to finding explanations for events or patterns of events. The mystery of so-called "dark matter" is a prominent contemporary example. Fritz Zwicky postulated its existence in 1934 to account for evidence of "missing mass" in the orbital velocities of galaxies in clusters. Subsequently, it has been invoked to account for the rotational speeds of galaxies, gravitational lensing of background objects by galaxy clusters, and the temperature distribution of hot gas in galaxies and clusters of galaxies. Astrophysicists want to know if such matter actually exists, just as nuclear physicists wanted to discover the Higgs Boson, so essential to the Standard Model. Explanations are more convincing when they include mechanisms seemingly responsible for observable events or patterns. In geology, Alfred Wegener proposed the theory of continental drift in 1912. It was based on the seeming physical match of the continents as pieces of a single jigsaw puzzle, but also on stratigraphic evidence indicating common morphology of rocks in places where continents were assumed to have once been joined. The scientific community rejected continental drift because it ran counter to the prevailing orthodoxy that the continents were fixed. Wegener also hurt his case by failing to offer a plausible mechanism to explain continental drift. The debate was reopened in the 1960s, partially due to additional evidence, but primarily in response to the emergence of a credible causal mechanism: thermodynamic processes deep within the earth that create convection currents that move the plates on which the continents rest.[38]

Scientists are attracted to thin, Humean, accounts of causation when no mechanisms or processes can be found to account for observable regularities. In the case of system-level theories, the Boltzmann equation or Maxwell transfer equations for rare gases are notable exceptions. More typical is the long-standing quest for kinetic equations for dense gases, which so far have failed to produce anything comparable. Physicists in fields where systems tend toward equilibrium also generally dispense with the concept of cause because it is not necessary to make accurate predictions. Instead, they search for statistical regularities, devise law-like statements based on them and subsume particular manifestations to these laws. A vocal minority objects to this approach, arguing that causation must somehow be brought into the picture.

Overall, physicists have dispensed with cause in two kinds of situations: when empirical findings violate conventional understandings of the relationship between cause and effect, as in quantum mechanics, and when attempts to find causes are unsuccessful but unnecessary for making reliable predictions. Physicists prefer, when possible, to bring cause into the picture. There is a long-standing belief— better described as an ideology—that explanations are an important component of knowledge. There is also a growing recognition in historical branches of physics, as in other historical sciences, of the importance of narrative causal accounts.

---

[38]Waldner, "Transforming Inferences into Explanations".

6.3   The Concept of Cause

With its emphasis on conceptual precision and logical consistency, philosophy has developed and interrogated diverse formulations of cause. The richness of this literature might paradoxically be considered its principal impediment. By demonstrating the inadequacy of all concepts of cause and ways in which they are incommensurable, philosophers of science unwittingly encourage us to reject as futile the project of developing a concept of cause that is logically rigorous and applicable to all kinds of events that might be considered causal.

The story begins with the pre-Socratic philosophers who sought to replace mythical accounts with empirical understandings of why things happened. Plato rejected these material philosophies and invoked an invisible world of forms to provide causes *(aitia)* for observable events. Aristotle offered a synthesis of earlier Greek thinking about cause that dominated Western thought down to early modern Europe. He distinguished four classes of cause *(aition)*. First, is material cause, out of which other things are made, as a statue is from marble or bronze. Material cause enables and constrains how matter can be shaped. Second, is formal cause, which is the form or template *(paradeigma)* for things, as an architectural drawing is for a building, or DNA for life. Third, is efficient cause, which is the primary source of change. Efficient causes can be natural or human. Fourth, is final end *(telos)* or the goal for which something is created. In the case of an acorn it is to produce an oak tree. To avoid an endless chain of regression, Aristotle argued against trying to trace any of his four types of cause back to their original causes. He was, however, interested in the concept of the unmoved mover. Every movement, he wrote, is due to some mover, but the first mover is unmoved.[39]

The medieval scholastics kept alive Aristotle's four-fold understanding of cause. Aquinas inserted god as the "unmoved mover" and creator of the universe. In the late sixteenth and early seventeenth centuries, many scientists and philosophers— distinctions between these fields were still in their infancy—sought causes in the natural world. Descartes, Malebranche, Hobbes, Gassendi, Boyle, Newton, and Locke were important figures in reducing and simplifying the number and kind of causes.[40] Hobbes was the most committed to a materialist philosophy. He treated cause as a two-body problem. Material substances interact in accord with the laws of motion. He attributed cause to the properties of the initiating agent and effects to the consequences of its action.[41]

To the extent there is a modern equivalent of Aristotle, it is Hume, whose formulation of cause is the starting point for contemporary understandings. Hume sought to purge the concept of all metaphysics and based it, as noted, on observable constant conjunctions. His Scottish colleague Thomas Reid was quick to observe that not all conjunctions are causal. Monday night always precedes Tuesday

---

[39]Aristotle, *Physics* 7.1–2, 8.4–5; *Metaphysics,* 2.2.

[40]Clatterbaugh, *Causation Debate in Modern Philosophy* and "Early Moderns".

[41]Hobbes, *De Corpore,* I.75–6, 102, 122.

morning but can hardly be considered its cause.[42] To distinguish between causal
and non-causal conjunctions, one must invoke some deeper understanding of cause,
which is precisely what Hume was trying to avoid.

John Stuart Mill attempted to salvage Hume's formulation by expanding it to
account for multiple causes in recognition that events are rarely the product of single
causes. Antecedents, whether singular or multiple, often require enabling conditions
to have their effects, and their absence can forestall them.[43] Mill developed his
methods of difference and agreement to tease out these conditions. His more
sophisticated approach to regularity builds on the intuition that a suite of conditions
might be necessary to produce a given effect. He nevertheless follows Hume in
insisting that there is no difference between cause and the antecedent condition of
any regularity; the former is the cause.

Mill's godson, Bertrand Russell, raised a different objection to regularities. He
observed that antecedent and consequent are assigned arbitrarily. What makes events
alike or different, and some of them members of the same class, require precise
criteria. These criteria can only be derived from an independent theoretical understanding of phenomena. This greatly complicates the problem of cause because it
requires a return to metaphysics.[44] Russell was untroubled because he insisted that
science had progressed beyond the concept of regularities to more functionalist
understandings. John Dewey made a similar point. "No event comes to us labelled
'cause' or 'effect'," he insisted. These are categories we invent and assign.[45]

Controversies about cause in philosophy revolve around three fundamental
questions: what is the nature of cause; is it a feature of nature or product of our
imagination; and is singular cause possible? How philosophers respond to these
questions generally turns on their understandings of the world and the relative
importance they assign to explanation and prediction. Their debates make frequent
use of examples, usually from physics, as most formulations of cause are intended
to apply to the hard sciences. Given this focus, there is often no difference in
moving forwards or backwards in time, making explanation and prediction
equivalent. Very little effort has been made to address social causes, where this
condition does not hold.[46]

What is the nature of cause? On this question, there is absolutely no consensus.
Humean and other regularity theories offer thin definitions of cause in an attempt to
avoid metaphysical foundations. Cause for them is efficient cause, and consists of
finding antecedents that consistently precede consequents. In some fields of physics

[42]Reid, *Essays on the Active Powers of Man*, Essay 4, Chap. 3.

[43]Mill, *System of Logic,* Book III, Chap. 5, Sect. 3, pp. 215–18; Psillos, *Causation & Explanation,*
pp. 59–66.

[44]Russell, "On the Notion of Cause".

[45]See Dewey and Bentley, "Knowing and the Known"; Davidson, "Action, Reason and Cause"
and "Causal Relations," for a recent statement of this argument.

[46]Chernoff, *Power of International Theory,* pp. 54–5; Jackson and Nexon, "Paradigmatic Faults in
International-Relations Theory." Notable exceptions include Nagel, *Structure of Science;* Hempel,
"Function of General Laws in History".

this works well because robust regularities can be found. It is also attractive in fields where physicists are unable to find deeper explanations for phenomena, as in statistical dynamics. The most serious limitation of efficient cause is that it depends on the discovery of regularities, if not constant conjunctions. Where they do not exist or are insufficiently robust, efficient cause is more difficult, if not impossible, to establish. Logically, the absence of constant conjunctions would compel us to conclude that there is no causation in these domains, which would be philosophically and conventionally unsatisfying. It also challenges a fundamental assumption of most regularity theorists that regularity and cause are features of the world.

Given these problems, many philosophers maintain that there must be more to cause than regularity. Some direct our attention to mechanisms or processes that might be responsible for events. The failure to secure a cargo door could explain an air crash, although a more satisfying explanation would also provide a reason, or reasons, why that door failed to close properly. We might also ask who was responsible for this problem: the aircraft designers or the ground mechanics. The mechanisms or causes we search for, and the level at which we search for them, are very much a function of how we frame a problem. One strategy to overcome the seeming superficiality of association is to subsume events or regularities to covering laws. The deductive-nomological (DN) model is the best-known statement of this position. Developed by Carl Hempel and Paul Oppenheim, it stipulates that an event (explanandum) is only explained when it is the product of a deductive argument, whose premise (explanans) is a law-like statement and a set of initial or antecedent conditions. These conditions need to be made explicit as must the links between explanans and explanandum.[47] In the plane crash example, the laws of aerodynamics could be invoked to explain why drag from an open cargo door prevented the aircraft from gaining the necessary speed and lift for a successful takeoff, although they could not explain why the cargo door was not properly closed.

The DN model cannot be described as causal because laws are not causes. It can, however, accommodate outcomes associated with observable statistical distributions. One of its principal drawbacks is that laws, by their very nature, are expected to be universal. This was not a problem for Hempel, who expected DN to be applied primarily to the physical world. It is, however, a serious impediment when we turn to the social world, where the level of scale is so often different. For physics and chemistry—the fields in which regularities in the form of constant conjunctions are possible—these regularities emerge from the interactions of phenomenally large numbers of atoms and molecules. One gram of hydrogen, for example contains 6×10^{23} atoms. These numbers allow scientists to describe system-level effects with a high degree of accuracy.

Only occasionally does this kind of analysis work in the social world. Engineers routinely treat all drivers and vehicles as equal when they model traffic flows, and assume that they all drive at the speed limit, which they know is not true in practice. They create closed systems and restrict their predictions to system-level effects. In

---

[47]Hempel, "Logic of Functional Analysis".

dealing with open and non-linear systems like international relations, this is impossible when the number of actors is small and the differences among them great, as is the case in international relations.[48]

Regularity theorists of all kinds advance 'thin' conceptions of cause. For them, correlations are as close as we come to cause. Many positivists insist that association is a first step in establishing cause, but sophisticated statisticians are the first to acknowledge that statistical techniques cannot discriminate between association and causation.[49] Rudolf Carnap insists that associations should be regarded as causal because they encode a "functional dependency of a certain sort."[50] Carnap and Schlick consider the concept of cause something of a distraction, as the goal of science is prediction. Carnap nevertheless thinks it essential to distinguish predictions based on laws of nature from those that are not.[51] For both philosophers of science, and for Hempel as well, prediction is the logical equivalent of explanation, a claim that has become known as explanation-prediction-symmetry thesis. Hempel acknowledges exceptions; statistical generalizations may account for past events but not predict future individual events. To address this situation, he developed the deductive-statistical (DS) model. He envisaged it as a placeholder for more complete DN account.

An alternate approach to cause goes back to Kant, who was troubled by Hume's empiricism and what he considered its likely social consequences. Hume maintained that experience leads to causality through inference. He regarded reason as a mere instrumentality. Kant thought Hume had reduced reason to "nothing but a bastard of the imagination fathered by experience." His goal was to show that cause, like all worthwhile metaphysical concepts, arises from "pure understanding." Causation is the form of understanding that authorizes inference of cause and effect. Kant is adamant that such inference is rule based. "Everything that happens," he writes in the *Critique of Pure Reason,* "presupposes something upon which it follows according to a rule."

Kant sought to provide a firmer grounding for cause by arguing that knowledge arises in space and time, intuitions about which are a priori categories of the mind. Cause cannot be established through Humean-style association because such claims lack strict universality or necessity. Nor can it be established through pure reason because it has the potential to produce dogma. We must explain events in terms of categories and associated concepts, and they do not arise from analysis but from intuitions that precede experience. To square this circle, Kant theorized—quite incorrectly, as it turned out—that our most fundamental intuitions are in accord

---

[48]Woodward and Hitchcock, "Explanatory Generalizations"; Lebow, *Cultural Theory of International Relations,* Chap. 1. See Byrne and Uprichard, "Useful Complex Causality," for a more upbeat take. They are optimistic about addressing causation in what they call situations of "restricted complexity," where it emerges from simple systems whose parameters are known.

[49]McKim and Turner, *Causality in Crisis?*

[50]Carnap, *Logical Structure of the World,* p. 264.

[51]Carnap, *Introduction to the Philosophy of Science,* p. 192; Schlick, "Kausalität in der gegenwärtigen Physik" and "Causation in Everyday Life and in Recent Science".

with nature. Absolute space, he insisted, is "not some adumbration or schema of the object, but only a certain law implanted in the mind by which it coordinates for itself the sense that arises from the presence of the object." Warrants of causality require transcendental justifications that subsume empirical observations to universal and absolute rules. Cause is a product of the mind but is empirically valuable because it describes the world.

Kantian causation looks beyond efficient cause to abstract 'noumenal' worlds to generate law-like statements that can be used to make sense of the empirical world. As noted, this approach is used in physics to explain equilibria and certain phase transitions. In economics and political science, rational choice arguably qualifies as Kantian when those who employ it recognize that their assumptions of *homo economicus*, or of leaders with clear and transient preferences, are artificial constructs that do not describe real actors. Like the transcendental worlds created by physicists to study non-equilibrium statistical mechanics, rational choice and rationalist models of politics and international relations incorporate assumptions that have no empirical referents.

Is cause inherent in nature or a human invention? Philosophers defend both positions. Ludwig Wittgenstein insisted: "The only necessity that exists is *logical* necessity."[52] Cause is a product of our imagination and has no scientific status. Hume also acknowledged the cognitive origins of cause, but, like so many other philosophers, desperately wanted to believe in a deterministic world.[53] This commitment gave an opening to Janet Broughton, Galen Strawson, John Wright, and other "New Humeans" to reinterpret the Scottish empiricist as a realist—unconvincingly in my opinion.[54]

The "New Humeans" treat cause as a feature of the world. Most go beyond Hume in insisting that causality must be regularity plus something else. Otherwise, it is impossible to distinguish between conjunctions that reflect laws of nature and those that are merely accidental or true by convention, like Reid's observation that Tuesday morning always follows Monday night.[55] Galen Strawson looks to forces to explain regularities, while other philosophers posit the existence of deeper laws of nature that are not themselves regularities. This 'thick' approach to cause is by no means limited to "New Humeans." It is appealing to a diverse range of regularity theorists, including Stephen Mumford, Brian Ellis, and George Molnar.[56] One formulation, known as the web-of-laws approach, originating with John Stuart Mill,

---

[52]Wittgenstein, *Tractatus Logico-Philosophicus*, §6.37.

[53]Hume, *Enquiry Concerning Human Understanding*, VIII.1.13.

[54]Strawson, *Secret Connection*, pp. 84–5; Broughton, "Hume's Ideas about Necessary Causation" and "Hume's Skepticism about Causal Inferences"; Wright, *Skeptical Realism of David Hume*; Beebee, *Hume on Causation*; Blackburn, "Hume and Thick Connexions"; Richman, "Debating the New Hume" and the other essays in Read and Richman, *New Hume Debate*.

[55]Kneale, *Probability and Induction*.

[56]Mumford, *Dispositions* and "Causal Powers and Capacities"; Ellis, *Scientific Essentialism*; Molnar, *Powers*.

was elaborated by F.P. Ramsey and subsequently by David Lewis.[57] It maintains that the only regularities that qualify as laws of nature are those that can be fitted into a larger, holistic structure. DN is consistent with this approach. Some advocates have attempted to demonstrate that laws of nature are necessary, an approach known as dispositional essentialism.[58] The principal objection to "regularity plus," DN included, is that it pushes back the search for causation to another ontological level. By doing so, Bertrand Russell long ago noted, it sets in motion an infinite regress of "unexplained explainers."[59] This problem prompted Wittgenstein to quip that "a nothing could serve just as well as a something about which nothing could be said."[60]

Regularity plus pits phenomenalists against realists. The former refuse to go beyond experience—that which can be observed and described—in their search for knowledge.[61] Embracing the phenomenalist position, some regularity theorists accordingly insist that regularities can only be explained in terms of deeper regularities, which would stay within the realm of the observable. These theorists acknowledge that any other kind of explanation would downgrade the importance of regularities.[62] Bas van Fraassen maintains that any account of laws needs to satisfy two conditions: it must stipulate how regularities are to be identified—that is, what distinguishes them from accidents—and how we make valid inferences from laws to regularities. Since laws can be considered regularities, reductive Humean accounts solve the inference problem but not the identification one.[63]

A middle position has recently emerged that attempts to build on Hume. Stathis Psillos argues that causation is a joint product of the world and humans. Likeness, he maintains, is based on "objective similarities and differences among events in the world and patterns of dependence among them." These likenesses are never exact but involve "degrees of similarity," and are thus, at least in part, "of our own devising." The designation of events and their categorization are acts of the human imagination, but can be made in response to "natural classes" of events that are "the product of the world alone."[64] In this account, cause ultimately has some objective reference, and if two researchers disagree one of them must be wrong.[65] Perhaps

---

[57]Mill, *System of Logic;* Ramsey, "Universals of Law and Fact"; Lewis, "Causation".

[58]Ellis, "Causal Powers and the Laws of Nature" and *Scientific Essentialism;* Ellis and Lierse, "Dispositional Essentialism".

[59]Hume, *Enquiry Concerning Human Understanding,* 4.17, offers the example of trying to explain bread in terms of its chemical composition. See Russell, "On the Notion of Cause," for the quote.

[60]Wittgenstein, *Philosophical Investigations,* §304; also Carnap, *Logical Structure of the World,* p. 264; Ayer, "What Is a Law of Nature?".

[61]van Fraassen, "To Save the Phenomena," *Scientific Image* and *Empirical Science,* for the views of a leading proponent of this approach. See Jackson, *Conduct of Enquiry in International Relations,* for a perceptive analysis of the two schools of thought.

[62]Psillos, "Regularity Theories".

[63]van Fraassen, *Laws and Symmetry,* pp. 38–9.

[64]Ibid.

[65]Williamson, "Probabilistic Theories".

this approach might prove useful in some branches of physics, but it is more difficult to apply to biology, where any typology of lifeforms is understood as subjective and to some degree arbitrary. In the social world where, individual human beings aside, there are no natural kinds, it is deeply problematic.

Other philosophers deny causality on empirical grounds. In 1909, prior to quantum mechanics, Franz Exner insisted that chance governed all natural events.[66] Ernst Mach and Karl Pearson made similar assertions in the first decade of the century.[67] In 1913, Bertrand Russell characterized causality as "a relic of a bygone age, surviving like the monarchy, only because it is erroneously supposed to do no harm."[68] The search for cause was dropped in many branches of science. The 1980s witnessed the growth in social science of artificial intelligence and Bayesian networks. Both approaches rely on the probability distributions and attempt to finesse the problem of cause.

Other philosophers have sought to widen, rather than narrow, our understanding of cause by moving away from Hume toward Kant, or even back to Aristotle. In 1932, Philip Frank contended that the law of causality could not be proven empirically but must be assumed a priori. He later modified his position to take into account Ludwig von Mises' conception of statistical laws. He insisted that it is impossible to derive any proof for the validity of a priori laws, but that science and everyday life depend on their existence.[69] Frank's position is a Kantian one as it distinguishes between an unobservable but nevertheless 'real' world that provides laws we use to describe the empirical world. In his 1957 magnum opus, he explicitly distinguishes between Humean and Kantian understandings of causation. He thinks the latter might provide the basis for reconciling causal and statistical laws.[70]

Another approach that has attracted considerable attention is causal process models. It is rooted in process philosophy, developed by Alfred North Whitehead, Paul Hartshorne, and Paul Weiss. Whitehead emphasizes the centrality of temporality, change, and passage to our world.[71] He insists that "Becoming is as important as being, change as stability."[72] Nicholas Rescher, the most prominent contemporary advocate of process philosophy, characterizes a process "as an actual or possible occurrence that consists of an integrated series of connected developments unfolding in programmatic coordination: an orchestrated series of occurrences that are systematically linked to one another either causally or functionally."[73] Process approaches are concerned with the problem of change and the dynamics or

---

[66]Lange, "Causation in Classical Mechanics".

[67]Mach, *Science of Mechanics*; Pearson, *Grammar of Science*.

[68]Russell, "On the Notion of Cause".

[69]Frank, *Causality and Its Limits*.

[70]Ibid., *Philosophy of Science*, pp. 238–9.

[71]Whitehead, *Concept of Nature*, Chap. 3.

[72]Rescher, *Process Philosophy*.

[73]Ibid., p. 22.

mechanisms that bring it about. The theory of evolution is the best-known process theory in science.

Among philosophers, Wesley Salmon is the leading theorist of causal process models. He builds on Russell's concept of "causal lines." Salmon defines causal processes as the "world-lines of objects exhibiting some characteristic essential for causation." A process describes anything with a relatively constant structure. It can be real or pseudo, as long as it leaves some kind of mark. The key is what Russell calls "quasi permanence," and Salmon reframes as "constancy of structure." A trajectory through time can be considered causal if it does not change too much and persists when isolated from other things. A ball moving through the air has quasi-permanence because it can be marked. A spot of light projected onto a wall does not transmit a mark, as changing the angle of the wall will change the shape of the light cone. Both are nevertheless continuous spatio-temporal processes.[74]

More distinctly Aristotelian is the causal powers approach. It is associated with scientific realism, a philosophy of science that received prominent statements in the works of Rom Harre and Roy Bhaskar and subsequent writings by Nancy Cartwright and C.B. Mart.[75] In international relations a variant known as critical realism is advanced by Heikki Potomaki, Colin Wight, and Milja Kurki, among others.[76] The core assumption of critical realism is that the world is largely independent of us and differentiated in the sense of being structured, layered, and complex. These layers interact, making all systems open and subject to emergent properties.[77]

Realists contend that the world should be understood as an ensemble of powers, propensities, and forces that explain complex interactions and events. Examples in physics include the spin, charge, mass, and decay of particles. Causal properties confer "dispositions on particulars that have to behave in certain ways when in the presence or absence of other particulars with causal properties of their own."[78] While causal properties are unobservable, scientific realists insist they are nevertheless a fundamental ontological category. Each causal property is expected to manifest itself in specific ways, but can remain latent if not stimulated, enabled, or otherwise activated. Causal properties give rise to tendencies rather than fixed laws (e.g., the tendency of states to balance against more powerful or threatening states), which, together with the open-ended nature of the world, explain why there are no

---

[74]Salmon, "Why Ask, 'Why?'," *Scientific Explanation and the Causal Structure of the World* and *Causality and Explanation;* Dowe, "Causal Process Theories"; Davidson, "Causal Relations" and "Laws and Cause".

[75]Harre, 'Powers'; Harre; and Madden, "Natural Powers and Powerful Natures" and *Causal Powers;* Bhaskar, *Realist Theory of Science;* Cartwright, *Dappled World* and *Hunting Causes and Using Them.*

[76]Patomaki, *After International Relations*; Kurki, *Causation in International Relations*; Wight, *Agents, Structures and International Relations.*

[77]Bhaskar, *Scientific Realism and Human Emancipation,* p. 61; Patomaki, "Concepts of 'Action,' 'Structure' and 'Power' in 'Critical Social Realism'".

[78]Bhaskar, *Possibility of Naturalism,* pp. 15–16; Chakravartty, *Metaphysics for Scientific Realism,* p. 108.

constant conjunctions.[79] Scientific and critical realists are interested in the kinds of causal properties that exist and the mechanisms that make them manifest.[80] This approach stands in sharp contrast to Humean association, in which the only relations that count are those of spatial and temporal contiguity. Causation is extrinsic to them and observed through cause-and-effect pairs of regularity.[81] Harre and Madden deride Humean association as "a picture of nature as a crowd of passive sufferers of external and imposed causality."[82]

Roy Bhaskar contends that most science operates in contexts in which the creation of closed systems, even in laboratories, is not feasible. This makes it difficult, if not impossible, to identify causal powers and generate laws.[83] His *Possibility of Naturalism* proposes that complex events be divided into their components, and that these be reframed as abstractions or idealizations. Rom Harre elaborates such a process of "iconic modeling," and offers it as a method to probe complex patterns of causation and eliminate weak hypotheses.[84] Heikki Patomaki considers it appropriate to evaluating all four categories of Aristotelian causation. Like other critical realists, he maintains that social relations are located in time and space, in an intransitive dimension that exists largely independently of our conceptions of the world or those of actors. Cognitive and discursive frameworks, which constitute the epistemological dimension of social relations, are nevertheless important.[85]

The most controversial causal power theorist is Nancy Cartwright, whose most recent books explore the implications of the causal powers for science. She argues that physical theories are rarely able to explain real events given the multiplicity of factors invariably responsible for them. We also require interpretative models and good engineering in all but the most carefully controlled laboratory experiments. Laws predict only when "nomological machines" have been constructed to guarantee that "all other things are equal."[86] She offers the example of a banknote falling from a tower. Newton's law of gravitation governs its motion, but does not predict the banknote's path or how soon it will reach the ground. This is because other forces come into play, most notably friction and air currents. Here, too, there are laws, but they are equally useless for purposes of prediction because they tell us

---

[79]Bhaskar, *Realist Theory of Science*, pp. 46–7.

[80]Mumford, "Causal Powers and Capacities"; Jackson, *Conduct of Inquiry in International Relations,* pp. 72–111, for an excellent overview.

[81]Beebee, "Causation and Observation".

[82]See Harre; and Madden, "Natural Powers and Powerful Natures" for quote, and *Causal Powers.*

[83]Bhaskar, *Realist Theory of Science*, pp. 46–7.

[84]Harre, *Philosophies of Science;* Harre and Madden, "Natural Powers and Powerful Natures" and "Exploring the Human Umwelt".

[85]Patomäki, "Telling Better Stories about International Relations" and *After International Relations,* Chap. 5.

[86]Cartwright, *Dappled World,* p. 50.

nothing about the specific wind vectors or their non-linear interactions. The broader point is that the laws of physics are abstract formulations that can rarely be applied directly to real-world situations without introducing multiple laws and lots of fancy engineering. The social world presents a greater challenge because we generally lack laboratories and the controlled variation they allow. In the absence of laws, which most often are derived from experiments, it is that much more difficult to identify the features of context that are most likely to affect particular processes and situations.

Physicists, she contends, must give up their cherished expectation that science will one day give us equations of unrestricted validity and applicability. Scientific truths exist at different levels and it is impossible to combine them into an overall, consistent formulation.[87] Some critics of scientific realism consider this approach enigmatic, if not quaint, given Cartwright's turn to Aristotle.[88] Others contend that the causal powers research program says little to nothing about the features of things that are causally relevant or the conditions in which latent powers become manifest. Nor, more importantly, does it tell us how we would even go about finding answers to these questions.[89]

Is singular cause possible? Singular cause refers to events that are causal but non-repetitive. Hume denies singular causation, as does the DN model.[90] Saul Kripke insists that "when the events *a* and *b* are considered by themselves alone, no causal notions are applicable."[91] Regularity theorists of all kinds consider singular causation something of an oxymoron.

John Dewey was an early proponent of what would become known as singular causation. He insisted that events are never repetitive as they are never the same in all their important dimensions. The social world is not mechanically reproductive because of agency, uniqueness and contingency.[92] Among contemporary philosophers, the most influential advocate of singular causation is Michael Scriven, who maintains that what has happened in the past, and what might happen in the future, is irrelevant to many causal sequences. He offers the example of a bottle of ink that spills and stains a carpet. Not all ink bottles spill, not all ink spills overflow desks, and not all desks are on carpets. We best account for such events by causal narratives that draw on situation-specific information. More general understandings may guide us to information relevant to these narratives.[93] For Humeans, causation is an extrinsic

---

[87]Cartwright, *How the Laws of Physics Lie; Nature's Capacities and Their Measurement, Dappled World* and *Hunting Causes and Using Them.*

[88]Giere, "Models, Metaphysics and Methodology".

[89]Woodward, *Making Things Happen,* p. 357.

[90]Hume, *Enquiry Concerning Human Understanding,* XII.138; Hempel, *Aspects of Scientific Explanation.*

[91]Kripke, *Wittgenstein on Rules and Private Language,* pp. 67–8.

[92]Dewey and Bentley, "Knowing and the Known".

[93]Scriven, "Explanations, Predictions and Laws".

feature of the world that takes the form of regularities. For singularity theorists, it is intrinsic and a function of the sequence of events connecting two relata.

Confluences, where a multiple stream of independent conditions come together to produce an event that would not otherwise occur, qualify as singular causation. J.L. Mackie offers the example of a house fire caused by short-circuit and a spark that ignites something combustible. Let us introduce further complexity by supposing that the owners left a candle burning on their windowsill—a common practice at Christmas time—and went off to a party. The window had a bad seal, allowing a draft to enter the room and blow the curtain close enough to the candle to catch fire. The fire alarm did not work because the couple had failed to replace a worn-out battery.[94] In either scenario, cause cannot be attributed to a single factor, and Mackie draws on Mill, to formulate the concept of INUS conditions. These are insufficient but non-redundant parts of a condition that is itself necessary but insufficient for the occurrence of the effect. Each of the contributing causes of the fire in the above example qualifies as an INUS condition. Mackie's approach allows us to account for complex causal situations, but is open to the criticism that it cannot distinguish between real causes and the cumulative effects of a possible common cause.[95] To do so, would require knowledge independent of and prior to the set of INUS conditions postulated to know that they met these conditions.

Mackie objects to Humean association on the grounds that it cannot encompass singular causal statements, but his account of causation is not that dissimilar from Mill's.[96] Donald Davidson moves even further in this direction than Mackie by insisting that singular causal narratives are reconcilable with Humean association if subsumed to some kind of law.[97] Anscombe has famously criticized this argument as representative of dogma unsupported by empirical examples.[98] Perhaps the strongest case for singular causation is made by C.J. Ducasse, who insists that objects and substances cannot be the relata of causal relations. The tree did not injure the lumberjack, but the falling tree did. Such events, Ducasse maintains, change the properties of things. In the examples he offers, there is no regularity. The next tree is unlikely to fall and the next one that does is unlikely to injure a lumberjack. Ducasse allows for causal regularities, but argues that Hume has improperly made them the norm and deduced causation from them. Regularities should be understood as 'corollaries' of causation, not definitions of them. Laws in turn are generalizations based on singular facts.[99]

---

[94] Mackie, *Cement of the Universe*, pp. 34–9.
[95] Psillos, *Causation & Explanation*, pp. 90–1; Kim, "Causes and Events".
[96] Mackie, *Cement of the Universe*; Psillos, *Causation & Explanation*, pp. 81–92.
[97] Davidson, "Causal Relations".
[98] Anscombe, "Causality and Determination".
[99] Ducasse, *Truth, Knowledge and Causation* and *Causation and Types of Necessity*.

6.4   Summing Up

Each of these approaches to causation has advantages and serious drawbacks. Humean conjunction is straightforward and builds on the seemingly hard-wired human cognitive proclivity to make causal connections between events. It finesses the need for thicker, metaphysical understandings, but only works in a limited range of circumstances where there are constant, or near-constant, conjunctions. It has problems distinguishing between regularities that reflect so-called laws of nature and those that are accidental. It cannot cope with singular causation. It is also debatable whether constant conjunction qualifies as causation because no mechanism responsible for it is specified.

Humean conjunction frames cause and effect as a binary relationship. This is a deeply problematic assumption, especially in the social world. The presence or absence of what are considered causes and effects, of necessity, involves arbitrary judgments. This is readily apparent when we look at the most commonly used definition of war by international relations (IR) theorists—interstate conflicts that produce 1,000 or more fatalities—or the definition of a recession in economics— two consecutive quarters of downturns in GDP. The bivalence principle of logic— which stipulates something is or is not the case—constitutes another conceptual drawback, as so many events cannot be categorized as "either-or." Hypothesized causes and outcomes in the social world usually come in degrees, many of them arrayed along multiple dimensions.[100] This is true for ethnic conflicts, interstate war, arms races, territorial conflicts, inflation, economic downturns, negative and positive balance of payments, loss of consumer confidence, and competition for scarce resources. How much of something is just as important a question as its presence or absence, as certain degrees of presence can produce phase transitions in politics, economics, and social relations more generally. Carl Hempel insists on a "high probability" of a statistical probability if something is to count as a cause. This criterion is woefully imprecise.[101]

Humean conjunction cannot account for equilibria. When simultaneous forces interact to maintain an equilibrium state, the study of regularities tells us nothing about this process.[102] Nor can regularities offer any insight into when a particular combination of forces will move a system out of equilibrium, or possibly into a new one. Humean conjunction fails to distinguish between difference making and production. Most of the time a bullet is fired into someone's head it proves fatal. Although the bullet produces death, it is not difference making if it was only one of a hail of bullets, as would be the case with a firing squad.[103] Even when a single bullet is fired and kills its intended victim, or a single sperm fertilizes an egg, these events represent only the last steps of causal chains. What we really want to know

---

[100]Kratochwil, "Of False Promises and Good Bets".

[101]Hempel, *Aspects of Scientific Explanation,* pp. 381–405.

[102]Martin, "Power for Realists".

[103]Godfrey-Smith, "Causal Pluralism".

are the conditions under which people are brought before firing squads and those in which they have unprotected sexual intercourse. All of these problems and omissions suggest that the concept of regularity does not map well on to the social world or tell us much that is critical when it does.

Regularity theories in science most often depend on large numbers and statistical probabilities. Physics and chemistry—the fields in which regularities in the form of constant conjunctions are possible—base these regularities on the interactions on phenomenally large numbers of atoms and molecules. They also establish, as far as possible, a closed system and limit their predictions to system-level effects. In dealing with open systems, this is not possible, especially in fields like international relations where the number of actors is small and the differences among them great.[104]

Regularity theories nevertheless appeal to many philosophers. Their defense rests on a foundation laid in the early nineteenth century by Scottish philosopher Thomas Brown. His *Inquiry into the Relation of Cause and Effect* dismissed the search for deeper causes as metaphysical and unscientific. Regularity is causation, pure and simple. Modern-day regularity theorists maintain that regularity has the additional advantage of providing the basis for prediction. They contend that their approach is consistent with realism on the grounds that regularities represent reality and are independent of the mind. As noted, this approach finds wide, but by no means universal, support among physicists.[105]

There are, of course, divisions within the regularity camp. Some philosophers subscribe to the so-called "strong formulation" that builds on Mill's response to Reid. It holds that it is not enough that 'A' regularly follows 'B,' but must do so under all conditions. Such a requirement is designed to make it unnecessary to specify mechanisms that might be responsible for the regularity.[106] Other theorists accept the probabilistic nature of regularities. Still others want to look for deeper regularities to explain the more superficial ones we find. They are all phenomenalists in their insistence that causation be limited to observables.

The DN model was invented to address the problem of causal versus accidental correlation by offering a more comprehensive conception of cause anchored in covering laws. Even more than its Humean predecessor, the DN model directs the attention of researchers away from explanation to prediction. Nelson Goodman observes that it reverses the stated relationship between a law and its implications: "rather than a sentence being used for prediction because it is a law, it is called a law because it is used for prediction."[107] Predictions may be possible in many branches of science but are notoriously unreliable in the social world for reasons I will examine in Chap. 2.

---

[104]Lebow, *Cultural Theory of International Relations,* Chap. 1.

[105]Psillos, "Regularity Theories".

[106]Ramsey, "Universals of Law and Fact"; Lewis, 'Causation'.

[107]Goodman, *Fact, Fiction and Forecast,* p. 21.

DN purports to describe the empirical world, but all social laws and the concepts on which they must be based are human creations. There is no privileged point outside the world from which to observe or engage it. Our concepts are the product of our language, a medium in which all researchers and human beings are immersed. Supporters of DN acknowledge that laws would be subjective, but insist they would not be arbitrary if they were repeatedly tested and refined by empirical application to the real world.[108] The DN approach—not very successfully in the opinion of its critics—attempts to walk a fine line between the perceived need to ground regularities while avoiding metaphysical commitments. To escape this dilemma, Willard Quine and Nelson Goodman suggest rooting law-like statements in natural kinds. Stathis Psillos observes that this approach verges on tautology "since it seems that for a regularity to be a law it should constitute a pattern among natural kinds... and conversely for a kind to be natural it should be part of a numerological pattern."[109]

Humean and DN models are universalistic in their claims, which is appealing to some researchers but unacceptable to constructivists. DN assumes a world amenable to covering laws. Such a world is not only realist but, more demanding still, must be hierarchical, with each ontological level sufficiently independent to allow the kind of parsimonious propositions on which covering laws are built. It would therefore be possible to integrate these propositions and laws into more holistic understandings. All these assumptions, Nancy Cartwright convincingly argues, are sharply at odds with empirical reality.[110]

Regularity theories, including DN, are realist because they treat causes and effects as natural kinds. Even if this were true, John Venn noted over a hundred years ago, they would still be subjective because they require reality to be organized in terms of event types.[111] Venn considered the construction of such categories a joint product of nature and human beings. Reference types could be constructed too narrowly or too broadly; they could exclude members of an event type or fail to exclude non-members. Regularity theories very much depend on properly calibrated descriptions of causal relata. This is often problematic in the sciences, and more so in the social world where none of the relata are natural kinds.

As noted, regularity theories and the DN model assume that repeated efforts at falsification will bring about a tighter fit between our theories and the world. In the physical and social sciences, refutations are rarely accepted as such but more often understood as problems of measurement or experimental error and "put right" by manipulating the data or explained away as anomalies.[112] On occasion, theories that are refuted subsequently turn out to be right, and even when they are wrong, may open up fruitful lines of inquiry. For different reasons, Quine, Kuhn, Lakatos, and

---

[108]Psillos, "Regularity Theories".

[109]Quine, "Natural Kinds"; Goodman, *Fact, Fiction and Forecast*.

[110]Cartwright, *How the Laws of Physics Lie*.

[111]Venn, *Principles of Empirical or Inductive Logic*, p. 98.

[112]Lebow, "What Can We Know?" [see also Chap. 2 in this volume].

Feyerabend have all argued that refutation is not only a difficult task but also rarely the reason why scientists give theories up.[113] So here, too, the practice of science differs notably from the assumptions about it built into causal models.

Scientific realism, the dominant causal powers approach, frames cause in terms of potential and enabling conditions. This two-stage formulation is appropriate to many physical and political phenomena. It has the further advantage of going beyond regularities in the search for causation. Capabilities can produce regularities, but only under the special conditions created by nomological machines.[114] Scientific realism's complexity may be its greatest drawback, as it is not evident how we determine the causal properties of things, or if this can be done independently of the conditions or processes that trigger or release them. For this reason, scientific realism skirts circularity. In Nancy Cartwright's example of aspirin, it provides relief in only some circumstances even if its capability will always be present.[115] But what kind of inference—if we exclude observable effects—would allow us to establish this universal capacity?[116] Realists maintain that causal properties of this kind can be established in laboratory experiments, which links realist explanations to a method of inquiry that has only limited applicability to the world of international relations.[117]

Kantian idealism also offers a richer formulation of causation that, in common with DN models, makes a fundamental distinction between association and cause. Cause must be something general, if not universal. For DN, there is a process of tacking back and forth between concepts and empirical findings, adjusting or revising the former to provide a better fit. For the Kant of the First Critique, we are part of the world and our concepts are often in tune with it by virtue of our nature, although they must also demonstrate empirical utility.

Kantian causation differs from DN in that its general concepts are more independent of the empirical world and never considered a superstructure to which empirical findings can be assimilated. Rather, they represent a noumenal world of abstract relations, which are unlike any empirical world but provide insight into it. In certain branches of physics, notably, statistical dynamics, this approach works well. In social science, Kantian causation finds its most direct expression in equilibrium theories—as long as their proponents consider them noumenal worlds that can be used to analyze empirical ones. In Chap. 4, I argue that noumenal worlds, like regularities, are useful starting points for causal narratives or forecasting.

---

[113]Quine, "Two Dogmas of Empiricism"; Kuhn, *Structure of Scientific Revolutions;* Lakatos, "Falsification and the Methodology of Scientific Research Programmes"; Feyerabend, *Against Method.* For a sophisticated counter-argument, see van Fraassen, *Empirical Stance.*

[114]Cartwright, *Nature's Capacities and Their Measurement*, p. 3.

[115]Ibid., p. 3.

[116]Morrison, "Capacities, Tendencies and the Problem of Singular Causes"; Psillos, *Causation & Explanation,* pp. 195–6.

[117]Wight, "They Shoot Dead Horses Don't They?"; Patomaki, *After International Relations.*

6.5 Lessons

My cursory overview of causation in physics and philosophy suggests several
conclusions. The first is the extent to which developments in philosophy parallel
those in physics. During the course of the twentieth century both disciplines were
drawn to diverse approaches to cause. Philosophers proved far more reluctant than
physicists to let go of the concept. There are discipline-specific reasons for their ·
perseverance, the principal one being the central place of logic in philosophy and
the possibility that cause can provide logical connections between events. At a more
fundamental level, the concept of cause offers a possible foundation for knowledge
claims.

The second and more important finding is the problematic nature of any concept
of cause.[118] Over the centuries, philosophers have devised different approaches to
cause in the hope of making it logically more rigorous and empirically more
applicable to the physical world. The most important step in this direction, which
began in the Renaissance, was restricting cause to one of its four Aristotelian
components: efficient, or immediate, cause. This radical move benefitted science,
but created as many problems as it finessed. Hume's formulation of efficient cause,
which offers an intuitively appealing simplification, is uncomfortably constricted
and incomplete in the absence of mechanisms linking cause to effect. The search for
mechanisms ineluctably drew philosophers and researchers back into the seeming
morass of multiple kinds of causes from which their predecessors had been desperate to escape. No formulation has been able to meet the test established by
philosophers of logical consistency and empirical adequacy.

The latter condition refers to its ability to account for all situations generally
considered to involve cause.

Social scientists have remained largely impervious to the developments in philosophy. Most pay no attention to questions of epistemology, and when asked to
defend what they do, tend to offer relatively unsophisticated or even confused
understandings. In the US, many social scientists fall back on some variant of
neopositivism. This is not without irony, as most consider the quest for cause the
goal of science but, in practice, restrict themselves to searching for associations.
King, Keohane, and Verba's, *Designing Social Inquiry* exemplifies both problems.

Physics has an important lesson to teach us: one size does not fit all. The
relevance of cause, and divergent understandings of it, differs across fields and
subfields of physics. Like physicists, we should feel free to approach causation in
ways that seem most appropriate to the subjects we study. Reasonable arguments
can be made for most other approaches to causation—especially if we treat them as
starting points for causal narratives that fold in idiosyncratic features of context. We
can nevertheless benefit enormously from clarification and critique of the ontological and epistemological assumptions and implications of these approaches to

---

[118]Psillos, "Regularity Theories".

cause. Recent books by Milja Kurki and Patrick Jackson make important contributions in this regard.[119] I have tried to follow their example in this chapter.

While an advocate of pluralism, I find existing approaches to causation unsatisfactory for reasons I have made clear. In the next chapter, I show that they are even more inapplicable to the field of international relations. I then outline my approach to the problem. It starts from the assumption that in the absence of constant conjunctions, we cannot finesse the problem of cause. We must look for the mechanisms and processes that may be responsible for the outcomes we observe. Even so, they will never do more than appear to explain some of the variation we observe. Mechanisms and processes operate in contexts, and these contexts determine the effects they have. We must consider two levels of context: the most superficial, where diverse conditions enable or constrain causal mechanisms; and a deeper level, where mechanisms themselves become possible. Some mechanisms are independent of cognition, while others depend on the understandings actors have of themselves and the world. We must on occasion go back to deep frames of reference that condition these understandings.

The intensive study of context draws us into what is unique about events and developments. This approach, that I call inefficient causation, is not to deny the use of event types; it would be impossible to do comparative analysis in their absence. Rather, it suggests that most events of interest to international relations scholars can only loosely be classified in terms of type because they are generally instances of singular causation. Events of this kind cannot be explained by general rules derived from past events; nor can their outcomes be used to predict future ones. Past events, and our understanding of them, are useful only to the extent that they become starting points for causal narratives that fold in context. General understandings can be for forecasting, but here too they do nothing more than provide initial entry points for narrative construction.

[119]Kurki, *Causation in International Relations*; Jackson, *Conduct of Inquiry in International Relations*.

Dartmouth College, N.H., USA

Founded in 1769, Dartmouth is a member of the Ivy League and consistently ranks among the world's greatest academic institutions. Home to a celebrated liberal arts curriculum and pioneering professional schools, Dartmouth has shaped the education landscape and prepared leaders through its inspirational learning experience. Dartmouth has forged a singular identity for combining its deep commitment to outstanding undergraduate liberal arts and graduate education with distinguished research and scholarship in the Arts and Sciences.

The charter establishing Dartmouth was signed in 1769, by John Wentworth, the Royal Governor of New Hampshire, establishing an institution to offer "the best means of education." Dartmouth's founder, the Rev. Eleazar Wheelock, a Congregational minister from Connecticut, established the College as an institution to educate Native Americans. Samson Occom, a Mohegan Indian and one of Wheelock's first students, was instrumental in raising the funds necessary to found the College. In 1972 it established one of the first Native American Programs in the country.

Dartmouth was the subject of a landmark U.S. Supreme Court case in 1819, Dartmouth College v. Woodward, in which the College prevailed against the State of New Hampshire, which sought to amend Dartmouth's charter. The case is considered to be one of the most important and formative documents in United States constitutional history, strengthening the Constitution's contract clause and thereby paving the way for American private institutions to conduct their affairs in accordance with their charters and without interference from the state.

Ranked No. 1 in undergraduate teaching for the last four consecutive years by *U.S. News and World Report* and recognized by the Carnegie Foundation as a "research

© The Author(s) 2016   117
R.N. Lebow (ed.), *Richard Ned Lebow: Major Texts on Methods and Philosophy of Science*, Pioneers in Arts, Humanities, Science, Engineering, Practice 3, DOI 10.1007/978-3-319-40027-3

university with very high research activity," Dartmouth combines elite academics with thriving research and scholarship.

For more than a quarter of a century, Dartmouth has hosted debates featuring presidential candidates. The College is a frequent stop on the campaign trail, giving students the chance to experience first-hand New Hampshire's first-in-the-nation presidential primary that every four years attracts candidates hoping to woo voters locally and capture attention nationally.

Dartmouth College educates the most promising students and prepares them for a lifetime of learning and of responsible leadership, through a faculty dedicated to teaching and the creation of knowledge.

- Dartmouth expects academic excellence and encourages independence of thought within a culture of collaboration.
- Dartmouth faculty are passionate about teaching our students and are at the forefront of their scholarly or creative work.
- Dartmouth embraces diversity with the knowledge that it significantly enhances the quality of a Dartmouth education.
- Dartmouth recruits and admits outstanding students from all backgrounds, regardless of their financial means.
- Dartmouth fosters lasting bonds among faculty, staff, and students, which encourage a culture of integrity, self-reliance, and collegiality and instill a sense of responsibility for each other and for the broader world.
- Dartmouth supports the vigorous and open debate of ideas within a community marked by mutual respect.

Since its founding in 1769 Dartmouth has provided an intimate and inspirational setting where talented faculty, students, and staff- contribute to the strength of an exciting academic community that cuts easily across disciplines. Dartmouth is home to about 4,200 undergraduates in the liberal arts and 1,900 graduate students in more than 25 advanced degree programs in the Arts and Sciences and at Dartmouth's professional schools: the *Geisel School of Medicine*, *Thayer School of Engineering*, and the *Tuck School of Business*. Dartmouth is also the first school in the world to offer a graduate degree in health care delivery science. Dartmouth faculty and student research contributes substantially to the expansion of human understanding.

Departments and Programs—Arts and Sciences

Learn and discover. At Dartmouth, education happens not only within traditional academic departments, but also at the intersections between them. Explore the 40 departments and interdisciplinary programs of the Faculty of Arts and Sciences.

Arts and Humanities	Interdisciplinary Programs
Department of Art History	African and African-American Studies Program
Department of Asian and Middle Eastern Languages and Literatures	Asian and Middle Eastern Studies Program
Department of Classics	Comparative Literature Program
Department of English	Environmental Studies Program
Department of Film and Media Studies	Institute for Writing and Rhetoric
Department of French and Italian	Jewish Studies Program
Department of German Studies	Latin American, Latino, and Caribbean Studies Program
Department of Music	
Department of Philosophy	Linguistics and Cognitive Science Program
Department of Religion	Mathematics and Social Sciences Program
Department of Russian	
Department of Spanish and Portuguese	Native American Studies Program
Department of Studio Art	Women's and Gender Studies Program
Department of Theater	

Sciences	Social Sciences
Department of Biological Sciences	Department of Anthropology
Department of Chemistry	Department of Economics
Department of Computer Science	Department of Education
Department of Earth Sciences	Department of Geography
Department of Engineering Sciences—Thayer School of Engineering	*Department of Government*
	Department of History
Environmental Studies Program	Department of Psychological and Brain Sciences
Department of Mathematics	
Department of Physics and Astronomy	Department of Sociology

Department of Government

Richard Ned Lebow is James O. Freedman Presidential Professor Emeritus; Professor of International Political Theory, Dept. of War Studies, King's College London; Bye-Fellow, Pembroke College, University of Cambridge.

King's College, London, UK

King's College London was founded by King George IV and the Duke of Wellington (then Prime Minister) in 1829 as a university college in the tradition of the Church of England. The University of London was established in 1836 with King's and University College London (UCL, founded in 1826) its two founding colleges.

It now welcomes staff and students of all faiths and beliefs. King's professors played a major part in nineteenth-century science and in extending higher education to women and working men through evening classes. The university has grown and developed through mergers with several institutions each with their own distinguished histories. These include:

- United Medical and Dental Schools of Guy's and St Thomas' Hospitals
- Chelsea College
- Queen Elizabeth College
- Institute of Psychiatry.

The staff and alumni of King's and its constituent institutions made major contributions to 19th-century science, medicine and public life, including Florence Nightingale. In the 20th century eight people from these institutions were awarded the Nobel Prize, among them Sir James Black, Desmond Tutu and Peter Higgs.

King's College London is dedicated to the advancement of knowledge, learning and understanding in the service of society. King's College London has a Faculty of Arts and Humanities, a Faculty of Life Sciences and Medicine, Faculty of Natural and Mathematical Sciences, Florence Nightingale Faculty of Nursing and Midwifery, Faculty of Social Science and Public Policy that include i.a. the Defence Studies Department, Institute of Middle Eastern Studies, Policy Institute at King's, Political Economy, *War Studies* and War Studies Online (distance learning). Furthermore King's college had in 2015 seven global institutes: African Leadership Centre, Brazil Institute, India Institute, Institute of North American Studies, International Development Institute, Lau China Institute and Russia Institute.

© The Author(s) 2016   121
R.N. Lebow (ed.), *Richard Ned Lebow: Major Texts on Methods and Philosophy of Science*, Pioneers in Arts, Humanities, Science, Engineering, Practice 3,
DOI 10.1007/978-3-319-40027-3

Department of War Studies

King's College established the Department of War Studies department in February 1962, with the first intake of students in September that year. The War Studies Group —comprising the departments of War Studies and Defence Studies—contributes to public life, participates in national and international networks, maintaining its international reputation for excellence in scholarship and policy-relevant research. The Department of War Studies is

- The only academic department in the world to focus solely on the complexities of conflict and security.
- Students are taught by experts and pioneers in their fields who are at the forefront of world events as they happen.
- Stellar academic cohort bring an extensive and continually growing network of national and international links around the world for students to take advantage of.
- Extensive range of events throughout the year hosting world leading speakers.
- Established relationships and links with major London institutions.
- Our location is close to government—physically as well as intellectually.

The Department of War Studies is committed

- To undertake and publish world-leading, cutting edge research.
- To provide outstanding, research-led teaching and training to the best students it can recruit.
- To disseminate knowledge generated within the Department through a range of knowledge transfer courses.
- To contribute to public life, participating in national and international networks, maintaining its international reputation for excellence in scholarship and policy-relevant research.

The Department of War Studies (DWS) is the largest European university group of scholars focused on research relating to all aspects of war, peace, security and international relations past, present and future seeking to

- produce world-leading research that develops new empirical knowledge, employs innovative theory, and addresses vital policy issues
- contribute to scholarly learning through high-quality publications, and to achieve impact through engagement and knowledge exchange with policymakers, parliamentarians, publics and industry in Britain and beyond.
- develop the next generation of scholars in international, policy, and security studies through postgraduate training and research mentoring
- support individual scholarship and research collaboration through excellent research resources and effective research mentoring;
- produce world-class scholarship through collaboration across the College and with international partners

Its *Impact Strategy* seeks to leverage the experience of colleagues with proven track records of achieving impact by sharing best practice with new research communities and early career researchers.

A cross-departmental research mentoring scheme is operated by the Department of War Studies, the Defence Studies Department, the Department for European and International Studies, the Department of Political Economy and the Global Institutes. This scheme allows for early career researchers to select a research mentor from outside their department within the Faculty of Social Science and Public Policy. It reflects the breadth and depth of academic experience located across the School and allows colleagues to engage with mentors that possess the most suitable research specialism and experience. Research mentors provide advice on research, writing, dissemination/publishing, funding, impact, networking and project design.

In the War Studies Department, *Richard Ned Lebow* is Professor of International Political Theory. He is teaching for BA students on Causes, Contigency and War and for MA students on Causation in International Relations, Politics and Ethics, Theories in IR, Concepts and Methods and Ancient Greek Conceptions of Order, Justice and War. See for more information at: http://www.kcl.ac.uk/sspp/departments/warstudies/people/professors/lebow.aspx.

University of Cambridge

The University of Cambridge is rich in history as one of the world's oldest universities and leading academic centres, and a self-governed community of scholars. In 2009, Cambridge celebrated its 800th anniversary. Its reputation for outstanding academic achievement is known world-wide and reflects the intellectual achievement of its students, as well as the world-class original research carried out by the staff of the University and the Colleges.

The reputation of Cambridge scientists had already been established in the late nineteenth century by Clerk Maxwell and the Darwins among others and was maintained afterwards by J.J. Thomson, Lord Rayleigh and Lord Rutherford. Work done by their pupils and associates during the Second World War greatly increased this reputation and large numbers of students flocked to the University and to government-sponsored institutions. University departments and research institutes were established as new areas of study developed. The 1950s and 1960s saw an unprecedented expansion of the University's teaching accommodation and the growing arts faculties received permanent accommodation for the first time.

The undergraduate numbers were increased after the war by the admission from 1947 of women students, by the foundation of a third women's College, New Hall (1954), as well as the foundation of Churchill (1960) and Robinson (1977). In the 1960s, four new Colleges were established for the growing number of teaching and research staff, as well as more places for research students. The older men's Colleges now began to admit women students and appoint women Fellows. Now 'co-residence' is usual, but three Colleges admit women students only—Newnham, New Hall, and Lucy Cavendish.

See at: http://www.cam.ac.uk/

© The Author(s) 2016
R.N. Lebow (ed.), *Richard Ned Lebow: Major Texts on Methods and Philosophy of Science*, Pioneers in Arts, Humanities, Science, Engineering, Practice 3, DOI 10.1007/978-3-319-40027-3

Pembroke College

Pembroke College, founded in 1347 by Marie de St Pol, Countess of Pembroke, is the third oldest of the Cambridge colleges. Openness characterises Pembroke today. The College is an intimate yet diverse community, committed to welcoming students of exceptional talent regardless of their social, cultural or educational background, and giving them the benefit of contact with a large and distinguished Fellowship. Pembroke thrives on conversations, between generations and disciplines—between undergraduates, graduates and senior Members, between current students and our alumni, and between the academy and the wider world.

At Pembroke College, there are around 440 undergraduate students studying for a degree at Pembroke. Pembroke also encourages the kind of inter-disciplinary discussions between staff and students in different subjects. The College is keen for its graduates to establish links with Fellows and other students in the same discipline and also offers graduates the opportunity of meeting people from other disciplines. At Pembrooke College, Professor Richard Ned Lebow has been a Bye-fellow in the field of international relations since 2011. For details see at: http://www.pem.cam.ac.uk/fellows-staff/fellows-2/bye-fellows/professor-ned-lebow/

© The Author(s) 2016
R.N. Lebow (ed.), *Richard Ned Lebow: Major Texts on Methods and Philosophy of Science*, Pioneers in Arts, Humanities, Science, Engineering, Practice 3,
DOI 10.1007/978-3-319-40027-3

About the Contributors

Bernstein, Steven (Canada) is Associate Chair (Graduate) and Director, Department of Political Science UTSG and Co-Director of the Environmental Governance Lab at the Munk School of Global Affairs, University of Toronto. His research spans the areas of global governance and institutions, global environmental politics, non-state forms of governance, international political economy, and internationalization of public policy. Publications include *Unsettled Legitimacy: Political Community, Power, and Authority in a Global Era* (co-edited, 2009); *Global Liberalism and Political Order: Toward a New Grand Compromise?* (co-edited, 2007); *A Globally Integrated Climate Policy for Canada* (co-edited, 2007) and *The Compromise of Liberal Environmentalism* (2001); as well as many articles in refereed academic journals, including *European Journal of International Relations, Science, Review of International Political Economy, Journal of International Economic Law, International Affairs, Canadian Journal of Political Science, Policy Sciences, Regulation and Governance,* and *Global Environmental Politics.* He was also a convening lead author and member of the *Global Forest Expert Panel on the International Forest Regime* and a consultant on institutional reform for the "Rio +20" UN Conference on Sustainable Development and its follow-up.

Address: Steven Bernstein, PhD, Canada), Director, Department of Political Science UTSG Munk School of Global Affairs, University of Toronto Mississauga, 3359 Mississauga Road,
Mississauga, ON L5L 1C6, Canada
Email: steven.bernstein@utoronto.ca
Website: http://www.utm.utoronto.ca/political-science/professor-steven-bernstein.

Gross Stein, Janice (Canada) is the Belzberg Professor of Conflict Management in the Department of Political Science and was the founding Director of the Munk School of Global Affairs at the University of Toronto (serving from 1998 to the end of 2014). She is a Fellow of the Royal Society of Canada and a member of the Order of Canada and the Order of Ontario. Her most recent publications include *Networks of Knowledge: Innovation in International Learning* (2000); *The Cult of Efficiency* (2001); and *Street Protests and Fantasy Parks* (2001). She is a

© The Author(s) 2016
R.N. Lebow (ed.), *Richard Ned Lebow: Major Texts on Methods and Philosophy of Science*, Pioneers in Arts, Humanities, Science, Engineering, Practice 3,
DOI 10.1007/978-3-319-40027-3

contributor to *Canada by Picasso* (2006) and the co-author of *The Unexpected War: Canada in Kandahar* (2007). She was the Massey Lecturer in 2001 and a Trudeau Fellow. She was awarded the Molson Prize by the Canada Council for an outstanding contribution by a social scientist to public debate. She is an Honorary Foreign Member of the American Academy of Arts and Sciences. She has been awarded Honorary Doctorate of Laws by the University of Alberta, the University of Cape Breton, McMaster University, and Hebrew University.

Address: Janice Gross Stein, PhD, Munk School of Global Affairs Belzberg Professor of Conflict Management, Department of Political Science, University of Toronto, 315 Bloor Street West (At the Observatory), Toronto, Ontario, M5S 0A7, Canada.
Email: j.stein@utoronto.ca.
Website: http://munkschool.utoronto.ca/profile/janice-stein/.

Weber, Steven (USA) is Professor at the University of California Berkeley, School of Information and of the Dept. of Political Science. He works at the intersection of technology markets, intellectual property regimes, and international politics. His research, teaching, and advisory work focus on the political economy of knowledge intensive industries, with special attention to health care, information technology, software, and global political economy issues relating to competitiveness. He is also a frequent contributor to scholarly and public debates on international politics and US foreign policy. One of the world's most expert practitioners of scenario planning, Weber has worked with over a hundred companies and government organizations to develop this discipline as a strategy planning tool. He went to medical school and did his Ph.D. in the political science department at Stanford. He served as special consultant to the president of the European Bank for Reconstruction and Development and has held academic fellowships with the Council on Foreign Relations and the Center for Advanced Study in the Behavioral Sciences, and was Director of the Institute of International Studies at UC Berkeley from 2003 to 2009. His books include T*he Success of Open Source* and most recently *The End of Arrogance: America in the Global Competition of Ideas* (with Bruce Jentleson) and *Deviant Globalization: Black Market Economy in the 21st Century* (with Jesse Goldhammer and Nils Gilman). He is working on a book, *Beyond the Globally Integrated Enterprise*, that explains how economic geography is evolving and the consequences for multinational organizations in the post financial crisis world.

Address: Steven Weber, PhD, Professor, University of California Berkeley, School of Information,102 South Hall #4600, Berkeley, CA 94720-4600, USA.
Email: stevew@ischool.
Website: http://www.ischool.berkeley.edu/people/faculty/stevenweber.

About the Author

Richard Ned Lebow (USA) is Professor of International Political Theory in the Department of War Studies, King's College London and James O. Freedman Presidential Professor Emeritus at Dartmouth College and also a Bye-Fellow of Pembroke College, University of Cambridge He has taught strategy and the National and Naval War Colleges and served as a scholar-in-residence in the Central Intelligence Agency during the Carter administration. He held visiting appointments at the University of Lund, Sciences Po, University of Cambridge, Austrian Diplomatic Academy, Vienna, London School of Economics and Political Science, Australian National University, University of California at Irvine, University of Milano, University of Munich and the Frankfurt Peace Research Institute. He has authored and edited 28 books and nearly 200 peer reviewed articles. Among his most important books are: *Franz Ferdinand Lives! A World Without World War 1* (Palgrave-Macmillan, 2014); *Constructing Cause in International Relations* (Cambridge University Press, 2014); (co-authored with Simon Reich: *Good-Bye Hegemony! Power and Influence in the Global System* (Princeton University Press, 2014); *The politics and ethics of identity: in search of ourselves* (Cambridge University Press, 2012); (co-ed. with Erskine, T.): *Tragedy and international relations* (Palgrave Macmillan, 2012); *Forbidden fruit: counterfactuals and international relations* (Princeton University Press); *Why nations fight: past and future motives for war* (Cambridge University Press, 2010); *A cultural theory of international relations* (Cambridge University Press, 2008); *Coercion, cooperation, and ethics in international relations* (Routledge, 2007).

Address: Prof. Richard Ned Lebow, PhD, Department of War Studies, King's College London, London WC2R 2LS, UK.
Email: ncdlebow@gmail.com.

© The Author(s) 2016
R.N. Lebow (ed.), *Richard Ned Lebow: Major Texts on Methods and Philosophy of Science*, Pioneers in Arts, Humanities, Science, Engineering, Practice 3, DOI 10.1007/978-3-319-40027-3

Websites: http://www.dartmouth.edu/∼nedlebow/;   http://www.kcl.ac.uk/sspp/
departments/warstudies/people/professors/lebow.aspx;   http://afes-press-books.de/
html/PAHSEP_Lebow.htm.

About this Book

In a career spanning six decades Richard Ned Lebow has made contributions to the study of international relations, political and intellectual history, motivational and social psychology, philosophy of science, and classics. He has authored, coauthored or edited 30 books and almost 250 peer reviewed articles. These four volumes are excerpts from this corpus. This second volume includes texts on causation, epistemology and methods, especially counterfactual analysis.

- This volume provides an overview of the research of a prominent scholar in the field of epistemology and research methods.
- It offers students of international relations and history a set of tools for using counterfactual experiments to probe complex causation.
- Here are many books in philosophy of science but none that directly compete. The author's approach is original and focused on international relations.

Table of Contents:

A book website with additional information on Richard Ned Lebow, including videos and his major book covers is at: http://afes-press-books.de/html/PAHSEP_Lebow.htm.

© The Author(s) 2016
R.N. Lebow (ed.), *Richard Ned Lebow: Major Texts on Methods and Philosophy of Science*, Pioneers in Arts, Humanities, Science, Engineering, Practice 3, DOI 10.1007/978-3-319-40027-3

Printed in the United States
By Bookmasters